Weathering
the Storm

Weathering
the Storm

Tornadoes, Television, and Turmoil

By Gary A. England

University of Oklahoma Press : Norman and London

England, Gary.
    Weathering the storm : tornadoes, television, and
turmoil / by Gary A. England.
        p. cm.
    Includes index.
    ISBN 0-8061-2823-2 (alk. paper)
    1. Tornado warning systems—United States.
    2. Television weathercasters—United States—Biography.
    3. England, Gary.
    I. Title.
    QC955.E54   1996
    551.5'092—dc20
    [B]   95-44997
                                                      CIP

Text design by Cathy Carney Imboden. Text typeface is
Janson Text.

The paper in this book meets the guidelines for
permanence and durability of the Committee on
Production Guidelines for Book Longevity of the Council
on Library Resources, Inc. ∞

1   2   3   4   5   6   7   8   9   10

To those who know me, it goes without saying that this is dedicated to Mary and Molly. Wife and daughter, their love is the driving force in my life.

This is also for Richard for setting the example, and Darla for trying to keep me in the light, and Phil for always believing in me.

And to Mom and Dad for bringing me into this world and giving me a chance to travel the road of my choice.

Contents

Illustrations

Maps

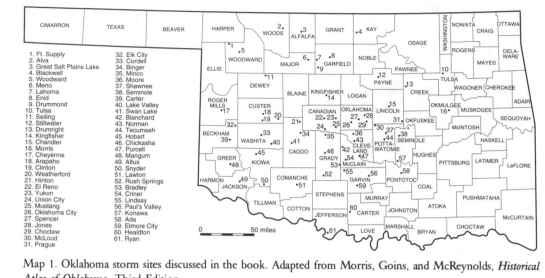

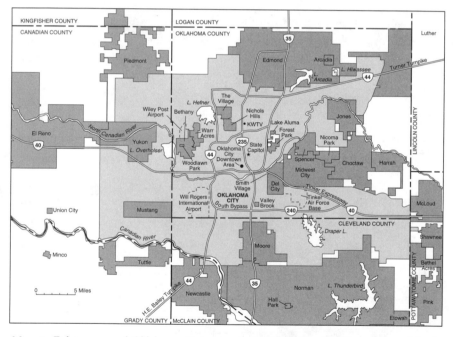

Map 1. Oklahoma storm sites discussed in the book. Adapted from Morris, Goins, and McReynolds, *Historical Atlas of Oklahoma*, Third Edition.

Map 2. Enlargement of Oklahoma County and surrounding area. Adapted from Morris, Goins, and McReynolds, *Historical Atlas of Oklahoma*, Third Edition.

Preface

"Priority one, tornado warning, take pointer over radar. I am ready!" Those words, urgent communications between myself and a television director, snap into motion an emergency weather warning sequence. Our sometimes frantic message flow can nearly be lost in the uproar that frequently engulfs the KWTV forecast center. Without fail, however, the system works. In moments I am on the air, live, delivering the urgent notification of an imminent threat to life and property.

This book is about such experiences, electrifying moments when humans face the awesome power of nature on the rampage. It is also a behind-the-scenes account of the hidden side of television—a turbulent world of conflict, humor, dedication, and competition. Corporate management, news directors, the government, the public, and meteorologists are all significant players.

These pages recount the decisions, forces, and struggles that have combined to produce the most highly regarded television weather market in the country. Although occasionally buffeted by criticism, both justified and not, television weather in Oklahoma City continues to set the standard for all others.

The events portrayed in this book are inescapably part of my life, and so this book is my story, too. Beginning with my boyhood dream of being a television meteorologist, this is the chronicle of my bumpy journey toward my personal perception of success in weather broadcasting. My path so far has been marked by thrills as well as disappointments.

Weathering
the Storm

Chapter 1

Lasting Impressions

A wagon pulled by two old mules creaked slowly down the street. Perched high on the springboard seat was a grizzled character out of the Wild West. His shaggy, gray beard partially hid a leathery face creased with lines of history. From under a dirty black hat pulled low on his brow, his intense eyes briefly caught sight of me.

I stood mesmerized as he and his rig made a methodical lumbering turn onto the side street. The mules, in what seemed like slow motion, edged up to an old, lone hitching post at the south side of the First National Bank. It was a hot, windy Saturday afternoon in 1947. I was back in Seiling, Oklahoma, the small, thriving town where I was born, and I was seeing things I had never seen before.

My family had just moved back to Seiling from Enid, Oklahoma, seventy-five miles to the northeast. In the short-grass country of western Oklahoma,

Seiling was a place where Indians still, on occasion, were shot without warning when caught breaking the law.

The servicemen had been home only a short time from the greatest war in history. It was commonplace for barroom brawls to spill out of the Terminal beer joint onto the street. Tommy Gun Ruble, a very colorful county attorney, could frequently be seen driving down the street on the way to help settle the more serious disputes, his famed weapon in hand.

It was illegal for Indians to buy alcohol, and liquor was taboo for whites. Bootleggers were extremely popular and busy. My Uncle John and Aunt Irma were among them. My mother, of course, always tried to convince me that John and Irma were very distant relatives.

The talk of the town then was the Woodward tornado of the spring just past, a powerful twister that had blasted through the nearby town on April 9 after dark, killing over one hundred persons. As a little boy, I was spellbound every time someone told about the horrors of that night. A large truck was never found. People had their clothes blown off their battered bodies. Lanny Sander, a lifetime resident of Seiling, told how his father, one of many volunteers who rushed to Woodward to help, found a naked person wrapped around a pole. Those were scary tales for any young kid to hear.

We were living in Enid when the storm struck. On the evening the tornado hit Woodward, the entire western horizon was covered with what looked like fuzzy, pink egg cartons floating gently across the darkening sky. All the neighbors gathered on the school ground and marveled at the eerie atmospheric display. I stood close to my mother, with pebble-laden sand squeezed between my toes and a piece of her dress locked in my hand. Before we turned to go home, my dad made the first weather forecast I remember hearing. In a matter-of-fact tone, he commented, "Somewhere tonight, there's going to be a bad tornado."

Safely in bed that night but still wide awake, I stared out the window. My lovable part-coyote, part-hound dog was, as usual, frantically digging holes in the front yard. Suddenly Cookie, his nose covered with dirt, became as rigid as a fence post. Moments later he began a low, mournful howl. Goose bumps sprang up over my entire body.

Within seconds, I heard the scream of sirens. For what seemed like hours, the neighborhood dogs howled. The sounds of sirens and the roaring engines that carried them pierced the night. Television was unheard of in our world and radio was mainly for entertainment. Therefore, we had no idea what was causing such a frightening exhibition.

Early the next morning, my dad came in and announced that the town of Woodward had been blown away during the night. The sirens we had heard were from ambulances, police cars, and fire trucks making the long run to Woodward in the hope of helping survivors.

The Woodward tornado and our time in Enid were definitely not on my mind, though, as that crusty character from the past dismounted his wagon. His mules stood motionless except for the rapid twitching of their ears and the swishing of their tails as they tried to rid themselves of pesky flies that swarmed about in random frenzy. The man stood there for a moment gazing at me. Intimidated by his presence I backed up against the screen door of McDaniels Grocery.

My fixation on him was interrupted when I became conscious of flies. Large, green, and aggressive, they swarmed around screen door. They swirled around my face—at least one attempted to fly right up my nose. But their main interest was the screen door and entrance into the grocery store.

The old man edged closer. With no change in expression, at last he broke the silence and gave me my second lesson in meteorology: "Boy, flies are always around the animals, but when they gather on the screen doors, it's gonna rain." Through the years, I have often seen that forecast technique beat the computers.

That evening, my family gathered, as we did every Saturday evening at the home of my grandparents, George and Stella Stong. The kids played cowboys and Indians, terrorized the younger children in the neighborhood, and on occasion behaved like total outlaws. The men played cards. They also consumed plenty of liquor.

Late in the evening, after the children had been forced to come into the house, the older adults would tell wonderful stories about the people and town. We loved to hear about Amos Chapman, a famous Indian scout who won the Congressional Medal of Honor at the battle of Buffalo Wallow. In the Texas Panhandle, he and four other men fought off over one hundred Kiowa and Comanche warriors. Amos was wounded six times and still managed to pull an injured comrade to safety. Later Congress passed a bill limiting the award to military personnel only. Therefore, the names of all Indian scouts who had won the award were removed from the list. That did not matter to us—Amos was still a hero. Even more exciting than the story of the battle was the rumor that his original log cabin was still somewhere just northeast of Seiling, on the North Canadian River. It was a mysterious place I secretly planned to find someday.

Gen. George Armstrong Custer quartered his troops at Camp Supply, just forty-five miles northwest of where Seiling was later founded. Custer and his troops savagely attacked and destroyed a large Cheyenne encampment about sixty miles southwest of Seiling, killing men, women, and children. Although never stated as fact, my uncle Ivan hinted that Custer's troops must have traveled very close to Seiling. Such information planted in our young and imaginative minds was unbearably exciting.

Ben Osage, who in his later years was an Indian scout, was rumored to have been in Custer's last battle when he was a young warrior. He died in 1937 at Seiling. It was thrilling to believe that a "wild" Indian, who must certainly have done in Custer all by himself, had ten years earlier walked the streets where we now played. I was bothered that his name was Osage because we were taught that only Sioux were in the battle. Later we were told that some Cheyenne were also involved. Ben Osage was a Cheyenne. Fact or fiction, we then firmly believed the tale.

In silent awe we heard stories about outlaws, including the Daltons, Bill Doolin, and the James gang who rode, robbed, and hid in western Oklahoma until near the turn of the century. Even the infamous outlaw Belle Starr was a part of our history. Old Ernie, a part-time chiropractor and a full-time storyteller, claimed that Starr once owned the only hotel still existing in Seiling. Ernie, though, claimed many things that even the children doubted.

In the 1890s, the Yeager gang had made their hideout just southwest of Seiling in the beautiful, rolling hills and canyons that gently wind their way down to the Canadian River. We were told about U.S. marshals who waited in ambush for the gang to come to the river for water and then, without warning, shot two of them dead, in the back.

Bonnie and Clyde used Oklahoma as a refuge from constant pursuit by the law. They were thought to be the outlaws who robbed the First National Bank in Seiling. The bank alarm in my uncle Earl's hardware store alerted everyone in the building that a robbery was underway. Uncle Earl ran out into the street, his shotgun in hand, just in time to see the villains' car roar away to the north in a cloud of exhaust and dust.

With somber delight, the ladies would tell tales about the evils of alcohol. They told us about Carry Nation. From 1894 to near the turn of the century, she made her home just two miles to the west of Seiling. To hear the ladies tell it, Carry, with her famous hatchet, chopped up every bar in Kansas and Oklahoma. The women suggested that if we grew up and behaved like our fathers and uncles, as well as other despicable individuals who frequented those devilish watering holes, we too would roast forever in the ovens of hell.

My great-grandparents, Isaac and Melissa, purchased the Nation homestead and lived there for many years. Just down the road from it, my dad would, time after time, point out the old buffalo wallows that then and now give silent testimony to the large herds of bison that once roamed this great land. Unbelievably, in 1945, the town of Seiling sold the Carry Nation house to Woodward as a tourist attraction. In 1947, that piece of history was blown away in the Woodward tornado.

Eventually on those wonderful Saturday nights, the subject would turn to weather, such as the terrible "cyclones" that swept the land clean of homes, trees, and people. Cyclones—that's what the old-timers called tornadoes. Time after time we heard stories about the same storms, usually with some variation but always fascinating. The Snyder cyclone of May 11, 1905, flattened most of that town. Eighty-seven people died, and scores more were injured. With very primitive medical care, the suffering endured was difficult to imagine. The dead were wrapped in blankets and stacked on flatbed wagons, pulled by horses. My uncle Aubry loved to describe the scene in intricate detail. His unbounded glee in frightening little children led to many sleepless nights. For years, anytime it thundered, I thought of those wagons, the blankets, and the people of Snyder.

During the dry times, we heard about the raging prairie fires that swept across the countryside on fierce Great Plains winds. Our minds held fearsome pictures of fields and homesteads left in charred ruins. Or conversation might center on the horrendous dust storms of the early and mid-1930s, which caused the great exodus of "Okies" to California. Tales of being nearly unable to stand up or see in the dark, dirty, powerful winds were popular. Grandpa Stong told about wrapping his face with cloth just to be able to breathe. It was frightening, but ever so fascinating.

Blizzard stories were always met with great excitement because all the kids loved snow. A great spring blizzard had swept the area with snowdrifts eight feet high in April 1938. That was my dream of what heaven would really be like.

My dad spoke many times of when he and his parents, three brothers, and two sisters lived ten miles southwest of Seiling. He would describe the snow-covered winter days when his dad would hitch up the horses, load all the kids on a large sled, and make the trek to school. How wonderful the sled trips sounded, but how rough they must really have been.

Dad, who quit school to help support the family, was an avid hunter. He told one particularly wild story about flash flooding. While on a springtime hunt at the nearly dry North Canadian River, he heard a distant roar suddenly grow louder. A few hundred yards upstream, he saw a wall of water filled with trees and debris churning toward him. Frantically he scrambled up a steep embankment and barely escaped as the raging water swept away everything in its path. He always related that story in a serious tone of caution.

Those tales of past and present weather experiences represented events we would encounter and have to overcome. We didn't think of them as safety precautions, but we were well trained in how to survive weather on the Great Plains. Never stand under a tree or pole during lightning, we were told, and don't touch the screen door. Stay away from creeks and rivers when it's raining upstream. Never swim in the river when it's rising or running fast, deep, and wide.

Convinced that our heritage required us to be tough and adventurous, many times we tested some of the instructions. I once felt compelled, on a beautiful spring day, to leap off a ten-foot bluff into a deep, wide, and swift river. The current carried me rapidly downstream along with several of my friends. After what seemed like eternity, I did reach the other side. I was terrified and shaking like an aspen. The trip back across the river seemed impossible. I had violated a rule that should not have been disregarded. In the process, I had learned to swim, but I would have given anything to have stayed on the shore.

Never drive into water running over the road was a firm rule. Always carry an extra container of water during hot weather trips, for yourself and the car in case it overheats. Wear a handkerchief over your face during dust storms or when caught in a blizzard or snowstorm. Also, when caught in a blizzard with a stalled car, always walk along the road or fence line and never wander out into an open field where you might lose all sense of direction.

My dad's advice was mostly proven to be excellent, but not always. "When a tornado approaches and there isn't a cellar available, go to the southwest corner of the house," he said. That turned out to be some really bad advice, as the southwest corner is usually the part of the house the tornado hits first. Another false recommendation was the procedure of opening windows in the house to equalize pressure in order to reduce tornado damage. We now know that the tornado opens the windows for you.

My favorite instruction of all was the one about heating bricks in the oven, then placing them on the floor of the car when making long trips during extremely cold weather. While a student at the University of Oklahoma, I had the opportunity to test that theory.

Temperatures were in the single digits when my wife, Mary, and I had to make a one-hundred-mile trip in a tiny car with a heater that blew only cold air. It was a bleak and brutal day. Frigid arctic air was screaming south across the gray landscape, destined for the warm coast of the Gulf of Mexico.

I placed two heated bricks on the floor of the passenger's side. I didn't want my little redheaded darling to suffer such a long, cold trip without some heat. The hot-brick theory proved correct. Mary's feet were fine, but a few days later, we discovered I had frostbite on both of my big toes. My dad usually knew what he was talking about.

In a region where life and livelihood were under constant assault from sudden weather changes and the associated ravages of nature, knowing what to do was critical. Knowing what the weather was going to do was essential as well. It seemed to me as a child that every adult was an expert in forecasting the weather. Some made projections on a continuous basis and were occasionally accurate. Others made dire predictions only when it was already obvious what kind of weather was on the way.

The thickness of animals' coats was a good indicator of the intensity of the coming winter. Spring and summer southwest winds suggested hot, dry conditions and a possible drought. East winds usually meant rain within twenty-four hours. Aches and pains foretold a change in the weather for the worse. When horses ran and kicked across the fields, thunderstorms would soon develop. When the birds ceased to sing, a storm was near. A ring around the sun or moon meant a change in the weather. Smoke from a campfire that rises and then moves away in a flat plane means fair weather. Dews and frost usually suggest nice weather. The list was endless and based only on the experiences of our parents and earlier generations. Many other weather indicators have been lost forever as new technology has replaced the signs of nature.

In 1947, radio weather forecasts were nearly nonexistent in western Oklahoma. Those that were available were considered the rambling of faceless fools who were always wrong. Radio reception was terrible. Interwoven with the distant voices was the constant, scratching roar of static. At least in my world, radios were few and far between. The one in our home had to be violently shaken every few minutes for it to maintain broadcasting any recognizable message.

Our family transportation was a big bread truck in which my dad delivered Martha Ann and Sonny Boy bread to a large section of western Oklahoma. The truck had no radio, so on those long, oppressive summer days riding on bumpy gravel or dirt roads, Dad would sing to me. "Old Number Nine" and "Oklahoma Hills" were my favorites. Had there been a radio, I would have been constantly searching each station for the latest weather information. I certainly wouldn't have those great memories of my dad singing to me.

So, for the most part, radio weather, and therefore professional weather information, was not a part of our lives. Like those who had come before us, we faced the wild and dangerous weather on our own, armed only with good advice from our family. We endured blistering heat and frigid cold. Incredibly strong, long-lasting winds blasted us from the north with snow and from the south with sand and dirt. Large, prickly tumbleweeds moving down the street at forty miles per hour were a common sight.

In 1947, television existed on the East Coast, but I doubt that anyone in western Oklahoma had heard of a device that would send pictures through the air. Such talk would have made a person the focus of unbearable jokes and abuse.

Tornado warnings usually came from officers of the law or on occasion from a resident west of town. As thunderstorms approached, members of the sheriff's office would place themselves a few miles west of Seiling, directly in front of the oncoming storms. Not a safe place to be, I might add.

If an observer saw an angry-looking cloud hanging toward the ground or an actual tornado, he would send word to Seiling by police radio. More often than not, the radio would fail and the storm spotter would be forced to drive at breakneck speed back to Seiling just ahead of the raging storm. With his siren blaring, he would head to the fire station, and within seconds the town siren would be alerting the town to the expected disaster. The warning was sounded nearly every time it stormed at night. For Seiling, it was almost always a false alarm.

Each spring storm season was anticipated with a certain degree of fear. Some of the most terrifying moments came when, over the roar of the storm, we heard the tornado siren. In the middle of the night, we would dash out the door, through the rain, hail, lightning, thunder, and wind, and pile into the damp, dark cellar. Aided by the dim, fluttering beam of a flashlight, we made a quick search of the floor and canning shelves for snakes and black widow spiders. The search was about as terrifying as the possible impending tornado. Soon the storm would pass, leaving the town intact but our nerves somewhat frayed.

One Saturday night at my grandparents' home, the adults were playing cards and we kids were outside. We saw distant lightning grow closer. Occasional thunder gradually became a continuous roar. Soon everyone was inside the house. The familiar wail of the tornado siren started, and preparation for the run to the cellar commenced instantly. The women quickly gathered the babies and rounded up the older children. We faced a fifty-yard charge across the yard and to the neighbors' cellar.

The roar of the wind and thunder was fierce. Blinding lightning flashed all around us. So bad was the storm, everyone in the house decided to go to the cellar—except for Grandpa Stong and Uncle Ivan, who insisted on remaining at the kitchen table, playing cards in hand.

The trek to the cellar was a momentous struggle. Leaning sharply into the mighty wind, the adults held the babies close to their chests, making sure their little faces were protected. Slightly older children were carried over an adult's shoulders like sacks of flour. My uncles Billy and Harold carried a child tucked under each arm. To see where we were going was impossible, so we moved in a line, holding on to the pants of the person in front. A barrage of rain and hail blew horizontally through the air, striking us like rocks shot from a cannon.

Finally, the group of frightened souls made it to the cellar, but the door was shut and tied down from the inside. As the clamor of the storm grew deafening, we beat on the door to alert those inside that others needed shelter from the storm. The door opened and we all tumbled down the steps to safety, joining the others already huddled inside.

There we waited for the tornado that surely was going to kill Grandpa and Uncle Ivan. The sounds outside were nothing short of terrifying. The noise of the storm momentarily subsided, and then we heard another brief burst of roaring and crashing approach and pass by. Finally, the sound of wind ceased and only thunder vibrated through the musty cellar air.

Expecting to find Grandpa and Ivan in the debris of the shattered house, we scrambled out of the cellar, with Grandma wailing and babies screaming, to face the awful truth. But to our delight, there beside the cellar door, covered from head to toe with mud, were the two tough guys who were going to ride out the storm in the house.

They sheepishly told how they had heard the sound of breaking lumber and metal swishing through the air, and then they had made a run for the cellar. They told us how they had pounded on the locked door and screamed at the top of their lungs for us to open it up, but the combination of the ferocious winds and screaming babies had made it impossible for any of us to hear them.

The small tornado that touched down just a short distance to the southwest of us destroyed only a shed and a few trees—and forever converted Grandpa and Ivan into true believers that one should always take shelter before a storm arrives. Of course, our experiences in dealing with storms were no different from others'.

Shortly after 10:00 P.M. on March 20, 1948, a large tornado, eight hundred yards wide, swept across Tinker Field in Midwest City, Oklahoma. In seven minutes the largest government repair and maintenance depot in the United States sustained approximately ten million dollars in damage.

As we struggled with the unpredictable weather, it never occurred to us that anyone would or could do anything about it. The disaster at Tinker Air Force Base, however, would bring about some truly great changes.

The day after that tornado, Maj. E. G. Fawbush and Capt. Robert C. Miller, two air force meteorologists stationed at Tinker, received their orders. They were directed by their commanding general to determine if tornadoes could be forecast. Relentlessly they reviewed the data, looking for a set of weather conditions that suggested the possibility of tornadoes.

Just four days later, on March 25, Fawbush and Miller were surprised to see a familiar weather pattern. It was very similar to the one that produced the devastating tornado that had struck Tinker.

In a following briefing, the general demanded a yes or no forecast. Fawbush and Miller responded, somewhat reluctantly, that they believed a tornado was likely to occur.

At approximately 6:00 P.M. that evening, a second tornado, estimated to be two hundred yards wide, smashed across the air base. Because of the forecast, the base was ready. Damage was kept to about six million dollars. The first tornado prediction in history, although not broadcast to the public, had been accurate. Those five days in March 1948 launched a turbulent period that resulted in the foundation of modern severe weather forecasting. Consistent success would be an elusive goal well into the distant future, but exciting new events were taking place in my world right then.

On a calm Saturday evening in early 1950, my dad told my mom, my sister, Darla, and I to get into the single, cramped seat in the bread truck. As we bounced, shook, and swayed down the rough streets, I wondered if Dad had possibly lost his mind. In a few moments, we came to a stop at the Seiling Hardware store.

As I looked through the large plate glass window, I saw many people standing in small groups around square, brown wood boxes. On the glass front of each box were faint images of people who appeared to be caught in a perpetual snow storm. Scampering inside, I quickly positioned myself front and center at one of those boxes. Uncle Earl proudly announced that we were looking at "television." It was nothing short of amazing. Moving pictures in a box. Finally, at 10:00 P.M., the hardware store closed. With strained eyes, we reluctantly departed.

Within a short time, people were buying television sets. As the town of Seiling is ninety miles from Oklahoma City, very tall antennas were required. Soon Uncle Earl and his crews were busy building television towers. The frail structures protruded fifty feet or more above the ground. The gangling antennas swayed violently every time there were strong winds. Many crashed to the ground, only to be rebuilt as quickly as possible.

I didn't have to worry about our antenna blowing down because we had no television. Therefore, I never missed an opportunity to stop by the hardware store to stare at and listen to what turned out to be my future. Late one Saturday night at Uncle Earl's store, I saw my first weather map. The lines on the map were intriguing. The low-pressure and high-pressure symbols on the mysterious chart were especially fascinating. That night, television weather captured me for a lifetime.

Grandpa and Grandma Stong were the first in the family to buy a television set. Unfortunately, my immediate family had moved to a farm a few miles west of Seiling, severely limiting my freedom to see television. But I had already discovered television weather, and it fascinated me just as much as the weather outside. And the weather outside was wild and varied.

The elements of nature we endured on the farm served to further heighten my interest. Summers were blistering hot. Nightfall, open windows, and a slight breeze offered the only relief. Strong fall winds on occasion created blinding dust storms. The dry soil from the surrounding wheat fields was torn loose and became a part of a huge, suffocating blanket of dirt. On bitter cold winter mornings, just stepping out into the howling northerly winds was a test of courage. Snow fell not down but sideways, it seemed, stinging exposed skin like a swarm of bees. Spring was a mix of beautiful sunrises, hot and humid afternoons, and violent thunderstorms.

One spring afternoon a fast moving thunderstorm surprised my dad and me as we worked near some farm outbuildings. The wind increased in velocity to near hurricane strength in a matter of seconds. With our vision obscured in the dirt- and debris-filled air, Dad literally dragged me to shelter in a large chicken house. We crawled on our hands and knees through the chicken droppings and, with all our might, held on to a supporting post inside the smelly, brick structure. For a brief moment, I opened my eyes just in time to see several chickens blow by like feathered bullets. The sound of breaking glass and tin ripping from the roof convinced me that the end of my life was near. Then, as is typical of Oklahoma weather, the storm ended quickly in one last flash of lightning and blast of thunder.

That storm terrified me, but for some reason, I also loved it. My dad commented at the time, "Good Lord, will we ever know when these darn things are going to hit?"

We didn't know it yet, but the art of forecasting severe weather had been progressing rapidly. An intense debate between the air force and the U.S. Weather Bureau (later to be known as the National Weather Service), however, slowed public access to severe weather forecasts. For fear of creating panic, the Weather Bureau refused to make public any tornado forecasts, although Major Fawbush and Captain Miller, continued to issue tornado forecasts for the air force. Those forecasts were the forerunners of what we today call tornado watches.

Finally, WKY-TV the only television station in Oklahoma City, was given access to the air force tornado forecasts. In the spring of 1952, Harry Volkman broadcast the first televised tornado forecast. It was well received by the public; there was no panic. It was a gutsy, intelligent, and farsighted move. I neither saw nor heard about the tornado forecast, but Harry Volkman had set the standard that would later guide my career in television weather.

Harry Volkman started his work in television at Tulsa, Oklahoma, in 1950. He was and is a genuine pioneer and leader in the field. Both professional and likable, he was the spark that lit an unquenchable fire in me to become a television meteorologist. From 1952 to 1955, Volkman broadcast television weather on WKY in Oklahoma City. By 1955, he was seen in homes all across Oklahoma via station KWTV. Harry moved to Chicago in 1959. There he quickly became as loved and respected as he had been in Oklahoma.

At least once a week, Mom, Dad, and I would get into the bread truck and go to my grandparents' house just so I could see Harry Volkman do the weather. Sitting cross-legged and far too close to the screen, I watched every movement he made and listened to every word he spoke.

One memorable Sunday evening, I pointed at Volkman and said to my father, "Daddy, I want to be one of those!"

He replied, "Well, what is he?"

To which I responded, "I don't know, but I want to be one!"

When Harry Volkman called for snow, I spent every waking minute waiting for the storm. Right up until bedtime, I would go to the door and stare at the distant, faint yard light in hopes of seeing at least one snowflake.

If the snow failed to fall during the evening, I would leap out of bed the next morning and run to a window. With breathless anticipation I would quickly wipe off the mix of water and ice from the inside of the window and desperately scan the yard for any hint of snow. If it wasn't there, no blame ever fell on Harry. My hero could do no wrong.

Chapter 2

Sailor Without
a Ship

Throughout the 1950s storms ravaged the landscape and disappeared over the horizon before television weather warnings could be broadcast. Residents of western Oklahoma discussed (and cussed) television meteorologists. Inaccurate warning bulletins were a favorite topic of conversation.

The onslaught of savage weather was not confined to Oklahoma. On May 11, 1953, a surprise tornado battered a large section of Waco, Texas. It destroyed large multistory buildings as well as homes. With 114 fatalities, it was the deadliest twister ever to hit Texas. The spring rampage continued. Less than a month later, on June 8, tornadoes killed 123 persons in Michigan. The following day, 90 persons died in Worchester County, Massachusetts, as a tornado that was three-quarters of a mile wide destroyed nearly everything in its path.

On May 25, 1955, unexpected tornadoes tore through Blackwell, Oklahoma, and Udall, Kansas, killing 21 in Oklahoma and 80 in Kansas. The southeast was not to escape the fury either. On April 15, 1956, violent tornadoes killed 25 in Jefferson County, Alabama. A well-documented, severe tornado blasted through Dallas, Texas, on April 2, 1957. That tornado, which left a track nearly fifteen miles long, resulted in 10 fatalities and a heightened awareness of how little was understood about severe storms. It also focused attention on the ineffective warning system then in use nationwide.

As population increased and communications across the nation improved, the public became more aware of the damage and death caused by the frequent surprise storms. Weather radars and severe storm evaluation techniques improved, but it would be many years before advance weather warnings would become a reality.

Before forecasting severe weather could become possible, additional knowledge had to be acquired about all the elements that eventually come together to create a thunderstorm. The first large, organized effort to understand this process was known as the Thunderstorm Project. Conducted during the late 1940s, it added to the limited knowledge of thunderstorm mechanisms. Smaller studies in the 1950s continued to investigate the complexities of the weather. Although these efforts provided a reasonable beginning, the road to a functional understanding of thunderstorm development and behavior was to be long and expensive.

Completely oblivious to the challenges faced in meteorology, I was certain that I wanted to be a television meteorologist. Oklahoma A&M, (now known as Oklahoma State University) offered the courses I needed. Following my graduation from high school, my older brother, Richard, drove from the University of Oklahoma, where he was completing his master's degree in geology, to Seiling and then, with me, on to Oklahoma A&M in Stillwater. Off we went, a young, naive dreamer and a brother who really cared. As we rode in his battered 1949 Ford down the narrow two-lane roads, Richard continuously talked about the importance of a college degree and the great future that could come with such an accomplishment.

About ten miles outside of Stillwater, with the hot summer air ripping through the open car windows, we passed a billboard. On it was a message that would soon make my mother sad, my brother a little upset, and my father happy: Join the Navy and See the World!

Two weeks after preenrolling at Oklahoma A&M, I found the message on that billboard too powerful to resist. I was fully delighted to join because of the navy's promise that I could attend the Navy Aerographer's Mate Weather School in Lakehurst, New Jersey. As I was only seventeen, my mother had to sign the papers that allowed me to serve my country, which she reluctantly did. The navy recruiter snatched up the documents like a hungry sea gull plucking a fish from the water.

With visions of the navy weather school and foreign shores bouncing around my head, I soon found myself on the sands of the Pacific Ocean in San Diego, California. Boot camp left no time to think about weather. Just making it through three months of a harsh, new world required all of my attention and effort. Besides, San Diego, in my opinion, had no weather. It was a beautiful place, but the weather was the same every day.

"You are ordered to report to the Naval Air Station at Beeville, Texas." Those were the words that greeted my graduation from boot camp. I was shocked because the recruiter had promised I would go to the navy weather school. Much to my relief, Beeville, Texas, would be only a three-month stay—and at least Beeville would have more weather than San Diego, but still nothing like the brawling, crashing storms in Oklahoma.

At Naval Air Station (NAS) Beeville, I was assigned to "chow hall" duty. My dream of becoming a meteorologist seemed totally submerged in a sea of potato peelings, tomatoes, radishes, and carrots. The daily effort of making salads for several hundred hungry sailors was not what I had in mind when I joined up. Therefore, in order to maintain a short-term goal and my sanity, covertly acquiring pies from the bakery became my primary objective in life. Eventually, with my five foot-seven-inch frame up to 196 pounds from its previous 150 pounds, my next set of orders came.

"You are ordered to report to the Aviation Training Base at Norman, Oklahoma." That directive was music to my ears. I learned that after thirty days in Norman, I would be transferred to the navy weather school in Lakehurst, New Jersey. Leaving NAS Beeville was a highlight in my life. As the wheels of that large navy aircraft left the ground, I felt human again.

Flying over Texas and then my beloved state toward Tinker Air Force Base, I closed my eyes and dreamed of what it would be like to study the weather. It was a great feeling.

I was jerked back to reality when the steady roar of the powerful engines was interrupted by loud, popping noises. A black plume of smoke pouring from one of the starboard engines caused me immediate concern. It was obvious there was a definite problem on how we were from going to get from where we were to the ground, in one piece.

Within seconds, a crewman appeared and unceremoniously threw a parachute at me. With a calm face and a booming voice, he shouted instructions on how to strap on the chute. Then, he told us, "Line up at the door! Jump! Count to ten and pull the rip cord!" At that instant I visualized myself tangled up on the tail of the aircraft, flopping around in the atmosphere that I would never have the chance to study.

As the twenty-four sailors and officers lined up for possible exit of the aircraft, I decided that there were not enough people on that plane to get me out the door. The flight was the third of my life. I had never seen a parachute before. Therefore, I was willing to take my chances with the aircraft. A tube of metal with three good engines sounded better than stepping out into the clear blue sky. Fortunately, about thirty minutes later we made a successful emergency landing at Tinker Air Force Base.

The old, faded blue navy bus that was to take us to Norman looked wonderful. I enjoyed every jarring bump in the road. Each confirmed that we were on solid ground. The atmosphere was a great place for thunderstorms, but not for humans, I had decided.

After thirty days of more marching, classes, and survival training, I left Tinker for my next destination: the Navy Aerographer's Mate Weather School in Lakehurst.

Traveling in a huge commercial airplane and just short of our destination, I asked the flight attendant, "What's that fluid streaming out of the wing?" With a sound that was a cross between a scream and a call for help, she ran toward the pilot's cabin faster than any person I ever met on the football field. Within moments the pilot announced that due to "a rapid loss of fuel," we would make an unscheduled stop at Washington, D.C. I had been enjoying the turbulence we had been flying through because it was the atmosphere in motion, but another aircraft emergency was almost too much to handle. I really wanted off the airplane. The military though, always gives you a no variance date and time of arrival. Finally, good judgment overruled panic. I remained on the aircraft.

The navy weather school was a dream come true. For the first time in my life, I was happy to carry a stack of books, go to class, and study. At orientation the chief petty officer gave his message loud and clear: for slackers, there would be quick termination; the top three students in the class would have their choice of duty stations. I set a conscious goal—to be one of the top three.

With high hopes and the kind of excitement that only youth can generate, I jumped into the program heart and soul. One class, though, was particularly confusing. It consisted of learning all of the rules for reporting weather observations. My adversary was a two-inch-thick government manual. Endless numbered subparagraphs graced the pages. Each time the instructor directed us to turn to a given location in the manual, I desperately looked for, but couldn't find, what he was talking about.

After two weeks, the instructor noticed that I was absolutely lost. "England! Do you have any idea what is going on?" Totally intimidated and barely able to speak, I confirmed his suspicions. On command, with quaking legs and rapid, shallow breathing, I handed my manual to him. After what seemed like eternity, he began to laugh. At this point I was sure the remainder of my navy career would be spent as a deckhand on some garbage scow plowing the waters of a dirty, long-forgotten harbor.

With my face hot, tingling, and surely fire engine red, the instructor explained the situation to me and the now-hysterical class. Whoever had used my manual previously must have shuffled the pages like a deck of cards. With no page numbers, just numbered paragraphs and subparagraphs, even the chief of naval operations could not have figured out where we were in the book.

The weeks at Lakehurst slipped by rapidly. It was like living a dream. Day after day, many of the weather mysteries that confused me and most everyone in Oklahoma were explained. With the Rocky Mountains close to the west and the Gulf of Mexico just to the south, western Oklahoma was well known as a zone of rapid, violent weather changes. Some storms moved in from great distances while others came to life right on top of us. I thought of how we Oklahomans didn't really know the difference or care. We just believed that no one except God knew what would happen next.

Sitting in class was like grabbing the brass ring. I eagerly learned that when cold air masses break loose from their icy origins in the polar regions of Canada and the Arctic, a massive movement of energy commences. Blocked by the majestic peaks of the Rockies on the west, the energy is focused toward the south. The leading edge of these air masses can move up to 60 MPH. From late fall to early spring, those cold outbreaks can turn a warm, calm, and beautiful day into an explosion of dirt and debris carried by air cold enough to penetrate any piece of warm clothing a person might wear.

Frequently, slower-moving cold masses of air will engage moist, tropical air from the Gulf of Mexico. The lighter air from the south glides gently up over the denser air from the north. This can create a range of weather that includes fog, drizzle, freezing drizzle, rain, freezing rain, sleet, snow, and thunderstorms.

From fall through spring, we learned, great rivers of wind high in the atmosphere frequently pass over the southern plains. The position of those winds, called jet streams, is determined by the varying tilt of the earth's axis.

The upper wind-flow patterns help to generate surface low-pressure areas. The lows can become a volatile mix of moisture from the Gulf of Mexico that is funneled northward by the Rockies and colder air from the north that is channeled southward, also due to the Rockies. Combined with blazing heat from the sun, that brew frequently becomes wild and turbulent anywhere east of the Rockies.

Transits of existing low-pressure areas from the west and over the mountains can cause rapidly changing weather conditions in Oklahoma and the other plains states. While over the Rockies, low-pressure rotations may become disorganized due to the high terrain and also due to the fact the vertical length of the rotation (from the bottom of the storm to the top) becomes shorter. The circulation is then stretched when the rotation moves down the east side of the Rockies to lower elevations. When the vertical length increases, the circulation center tends to become smaller. This can decrease air pressure, increase wind speeds, and produce rapidly rising air. The results can be storms that appear to develop without warning. I could hardly wait to get home and impress everyone with my new-found knowledge.

With mixed emotions, weather school came to an end. An old, craggy chief petty officer with the personality of a pit bull came to announce our next assignments. I was really excited. Having placed among the top three of my class, I expected to have my choice of duty stations. My preferences were all aircraft carriers, all in the Pacific command.

As the chief barked out names and assignments, I heard yells of joy mixed with groans of disappointment. "England, Naval Auxiliary Air Station. Fallon, Nevada!" With those words the weight of the world came crashing down on me. I slumped to the floor, a dispirited, depressed person, a sailor without a ship or an ocean. Placing third in my class earned me a year on a sea of sand, not my choice of duty station, as promised.

Naval Auxiliary Air Station (NAAS) Fallon, about sixty miles east of Reno, is located in a desolate, bleak, virtual desert. It looked to me like the end of the world. With great reluctance, I reported for a year of duty and was immediately assigned to X Division for thirty days. There I languished, swabbing landlocked decks and policing the area by picking up trash around the barracks. Television weather was, for me, feeling like the impossible dream.

Occasionally I would work a late night or early morning shift for the Officer of the Day. There I carried out such duties as making coffee, answering the phone, and running up the flag. One beautiful morning, and there weren't many, I ran up the flag as the scratchy bugle recording split the crisp morning air. About halfway up, the lines became tangled. Old Glory was stuck, and I panicked. Tying the proper knot, I turned smartly and returned to the safety of the office. With the flag flying at half-mast, I was praying my shift would end before someone noticed.

"Hey chief! Who died?" With those words I broke into a cold sweat. Timidly I peaked around the door and there stood a stiff and well-starched lieutenant commander, talking to the chief petty officer of the watch. The chief took on the appearance of a human bomb with a very short fuse. His face was a bright crimson. The tongue-lashing I received seared my brain like a red-hot cattle brand. Within two days, however, I was transferred to the weather office. That was the first time I realized that good things can come from bad experiences.

Finally being in a position to use what I had learned in weather school was fabulous. The weather chief petty officer was as mean as a junk yard dog, but the joy of working in an office with real weather instruments and other weather personnel outshone his vicious behavior.

Taking hourly weather observations, plotting maps, and determining upper winds are the most basic duties of forecasting, but I found the work nothing short of fantastic. Measuring what the atmosphere was doing flipped on my switch again. The dream of television weather returned to my conscious thoughts.

My year of living and working in the middle of a desert was not rewarded. An anonymous personnel officer selected me to report to the middle of the Pacific Ocean—Midway Island, twelve hundred miles northwest of Hawaii. Now I would spend a year of duty on a ship made of sand. Home was beginning to sound pretty good.

As I settled into a cramped canvas seat, my back against the rough fuselage of a Military Air Transport Service aircraft, I was beginning to think that I'd been had. I would forever be a sailor without a ship, for this would be my last duty station.

The crewman who instructed us on safety procedures cut right to the chase. "If we crash into the water on takeoff, your life preserver is under your seat. If you are still alive after impact, exit the aircraft in an orderly manner and await rescue. Keep your senses about you and remember, you have been trained to swim in water that is covered with burning fuel." After those comments I tuned him out and tried to think of home and of Harry Volkman and his television weather shows.

With roaring engines and vibrating metal, the aircraft lifted its cargo of young, homesick sailors from the runway. After a while I saw the lights of San Francisco fade rapidly away in the pitch black sky. I then realized that I was about to spend a year on an island that was about a half-mile wide and a mile and a half long. It was inhabited by five thousand sailors and Marines, and tens of thousands of albatross, commonly known as "gooney birds."

Midway turned out to be an enchanting bit of sand. Several species of birds made their homes in the tall, beautiful trees. The tiny, white, fairy terns were exceptionally tame and would sometimes lightly balance on a friendly shoulder. When the large gooney birds returned after months at sea, those masters of flight would glide in over the crystal blue ocean waters. As the waves rolled over the fine sand, bleached nearly white by the intense sun, the albatross rolled, tumbled, and crashed into the island. It was quite a sight. They were used to landing in the water, and while at sea each year apparently forgot their ground-landing techniques.

On hot days with little or no wind, the large, ungainly birds would run, wings flapping wildly, long distances in vain attempts to become airborne. For this activity they used the beach, but they preferred the aircraft runways. Frequently, planes had to abort landings and takeoffs because of the large number of feathered friends running down the long concrete slabs. The hotter the day, the further they had to run. It was a real-life lesson in meteorology. The higher the temperature, the less dense the atmosphere, and therefore the longer it takes a gooney (or an aircraft) to get off the ground.

Working in support of the aircraft and crews flying the "radar barrier" between Midway and the Aleutian Islands was serious business as well as solid, heavy-duty weather experience. The weather they flew through— around the clock—was some of the most violent to be found anywhere in the world. Their mission, in radar-laden Super Constellation aircraft, was to give the first warning of any Russian attack originating from the northwest. For the air crews, just surviving the weather was a major accomplishment.

Gathering data and helping to forecast weather for the unfamiliar climate to our north was a significant challenge for all involved. Fast-moving, low-pressure-area storms extended from the ocean surface to well above flight levels. Visibility was frequently zero, and wind velocities were extreme.

We worked a shift continuously for nine days before having three days off. This provided a wealth of knowledge on dealing with fast-changing, life-threatening weather. Data had to be collected, and had to be correct. The data had to be analyzed accurately. There was no time to hesitate. Decisions had to be made, and they had to be right. It was superb training for the job I would hold in the distant future.

I was first exposed to weather radar on Midway. Just being in the darkened room with that green-tinted circular scope was so exciting that I shook. I could actually see the storms the "Super Connies" were flying through. The radar room became my second home.

As the year on Midway drew to a close so did my tour in the navy. I had no mixed emotions about leaving. The meteorology experience and training had been superb on the island, but home is where the heart is.

Departing Midway, I looked out the aircraft window. Far below, little cumulus clouds were dotted over the deepening blue of the Pacific. The two-engine aircraft suddenly shuddered. It then banked steeply to the right. A knot in my stomach tightened. I closed my eyes and prayed. I silently informed whoever might be listening that I was a sailor and should be on a ship, not violating the pristine atmosphere. The forty-minute return flight to Midway was a hair-raising experience. I decided then and there that I would study the workings of the atmosphere from the ground.

Chapter 3

Mary and Camille

With the navy fading quietly into the past, I began school at Southwestern State University at Weatherford, Oklahoma. It was the fall of 1960. Enrolling in sixteen credit hours of math, physics, English, and other courses, I was bulging with pride and filled with hope for the future. Unfortunately, after spending a year on an island, my mind was not on school.

I could find no time to study, but I had plenty of time for all other activities. One evening while standing between two cars talking to a young woman with whom I had a date, I spotted a cute little redhead coming my way. As she passed within arm's length, I reached out and pulled her nose to nose with me. For some unknown reason, and with no warning, I kissed her. She ran, jumped in a car with her friends, and sped away. My date rewarded me with a slap in the face. That was how I met Mary Helen Carlisle, who a year later would be my bride.

As I became more focused on fun, classwork became even less important. On finishing my first semester, my advisor told me that if I did not start behaving and studying, I would be fortunate to remain a student, let alone ever becoming a meteorologist. I had dropped all but seven credit hours during the semester; the future looked bleak. My dad wasn't impressed with the one A I received. "Folk dancing has nothing to do with meteorology!" he snorted. I, therefore, spent the spring semester out of school, driving tractors and cutting wheat.

The next fall I finally arrived at the University of Oklahoma in Norman. Mary and I had gotten married. My dream of doing television weather was intact and growing stronger each day, but little did I realize how much it would take to realize that dream.

"You'll never make it," said my advisor after noting that I had taken only one math course in high school and had single-handedly disrupted a good part of campus life at Southwestern State. "Meteorology is math and physics," he mumbled, "but go ahead and try. You can change later to something you can handle."

The University of Oklahoma meteorology program was struggling through serious growing pains about which I knew little. To obtain what today is a degree in meteorology, one had to choose a major in math, physics, or engineering. Then one could take all of the appropriate meteorology classes and receive what was called a "meteorology option." With great reluctance I decided to major in mathematics with a meteorology option.

Starting with college algebra, I stumbled, crawled, and fought my way through what seemed like a dark, endless tunnel of math courses. To meet the foreign language requirement, I started out in Spanish and then switched to German, believing it would be easier. I soon went back to Spanish. The elective courses, I felt at the time, were a bother and of little value to a future television meteorologist.

After one really nice semester of just going to school, I came home, and, with no hesitation, my wife announced that she had found me a job. I really was not interested in having a job, so I was less than pleased. Directing me with the firmness of a military commander, she said I was to report to the the Atmospheric Research Lab on the north campus. There I would be interviewed by Dr. Walter Saucier.

"Doctor who?" I asked, trying to contain my anger and fear that my easy life was about to end.

"Dr. Saucier, 2:00 P.M. tomorrow!" came the firm answer.

Dr. Walter Saucier, a tall, dynamic, and brilliant man with a very dry sense of humor, was fascinating. After an hour of lecturing, pacing the floor, and verbally deriving equations that I would not face for another two years, he offered me the job. I was confused by the interview but excited. My wife, on hearing the news, was nearly overwhelmed with the mere thought that I had a real job.

Dr. Saucier assigned me to work for Victor Whitehead, a thin, soft-spoken intellectual with a great sense of humor. He was doing research on the upper atmosphere for his doctorate in meteorology.

Whitehead maintained his cool even in the worst of times, like when I accidentally dropped several hundred computer punch cards just after he had told me not to drop them. When I had no money, Whitehead would assign extra work for me to do. If I had difficulty with a class assignment, he was always available to help.

Working with leading-edge atmospheric scientists who were just beginning to investigate the great mysteries of the atmosphere provided an education as valuable as college. They were revealing the secrets of vortex generation, the lightning factor in tornadoes, and numerical forecasting techniques. They developed new approaches to projecting the future location of air parcels flowing along in the great river of air that encircles our earth. Those were just a few of the activities that made life so exciting.

The Rough Rider project, which used aircraft to study storms at close range, was like a magnet to those of us fascinated by meteorology. When one of the aircraft was penetrating a thunderstorm to gather vital data on storm structure, a group of us would always be gathered around the radio room. We listened intently to the pilot's comments as he made his violent ride through one of nature's more magnificent creations.

In 1963, while the Rough Rider project was underway, the U.S. Weather Bureau established the National Severe Storms Laboratory (NSSL) at Norman, Oklahoma. This was a stroke of genius. Through the years to come, the NSSL would reveal the secrets of the structure and behavior of severe storms. Doppler radar, which measures the movement and speed of precipitation particles, would be tested there and proven to be the ultimate tool for severe storm detection and evaluation. The NSSL would greatly advance the transfer of research technology to operational meteorology.

Walking down the old, uneven, wooden halls at work one day, I passed a person with a very familiar face. He nodded hello, and I froze. Eyes wide and mouth open, I could not manage an intelligent sound. Whitehead noticed my panic and asked, "How would you like to meet David Grant?"

"You bet!" I blurted, and continued walking the other way.

David Grant was the meteorologist who had replaced Harry Volkman at KWTV. He was a bright, vigorous former air force officer. Whitehead took me by the arm and gently turned me around. Face to face with a real television meteorologist, my fear intensified, but Grant's warm smile and gift for gab calmed me down somewhat. After a brief conversation in which my comments made little or no sense, Grant invited me to visit him at the television station. I felt like I had just met the Pope.

Grant's main competition was Bob Thomas at WKY-TV. Thomas, an excellent meteorologist, had gained considerable fame in Oklahoma. His weather was sponsored by a beer company. Their commercials were animated cartoons and were very humorous. Every time one of those commercials ended and the camera came back to Thomas, he would be laughing, tears streaming down his normally straight face and nearly unable to continue his weathercast. The audience, myself included, loved it. I watched Thomas and Grant every chance I had, but school was the focal point of my life.

College was difficult. Each semester harbored monstrous surprises that seemed to leap out from dark corners and threaten my chosen path to success. Only years later I would find that dyslexia was my hidden opponent.

Many times my senior year, I visited Grant at KWTV. Being a gracious person, he always took time to talk with me and offer encouragement. He once mentioned that a weekend on-air meteorologist position might become available during the upcoming summer. When he made that statement, I became as nervous as I was excited; but I was graduating that spring, and it was time to dance or get off the floor.

Dreaming about being on television was easy. Practicing the weather in front of a bedroom mirror, as I had done for years, was also easy. The actual thought, however, of standing in front of a real television camera, with real people watching was an alarming idea.

At the urging of my wife and Grant, I agreed to make an audition tape. Getting me to do it was like trying to get a pig to go where you want it to go. When the day of the appointment arrived, I forced myself to show up, but to this day I have no idea what I said or did during the audition. I do recall, though, that I felt very hot and damp when it was over.

I thought I had no chance of being offered the weekend weather position at KWTV, so I proceeded with other plans to make a living. Jerry Osborne, a thin, quiet, and pleasant individual working on his master's degree in meteorology was involved in those plans. We had convinced ourselves that we could establish Oklahoma's first commercial weather service, sell weather information, and become rich!

The offer from David Grant for me to do weather on the weekends at KWTV came like a bolt out of the blue. As though I had lost my mind,

I turned the offer down. I had a chance at my lifetime dream, and I said no. It took me years to figure out that fear of actually being on television caused that decision.

Osborne and I continued with what became Southwestern Weather Service. Selling weather information to aviation, construction, and agricultural interests and teaching weather ground school to student pilots, we pursued the elusive riches of being in business for ourselves.

Our primary client was Shamrock Airlines. It maintained offices upstairs from us and mainly flew the Oklahoma City Blazers hockey team to their matches. Flying in good weather and bad, their planes always made it back, probably because of the skill of the pilots rather than our forecasting.

After two years with Southwestern Weather Service, my wife announced that we would be having a baby. I had mixed emotions—great happiness about the baby, but considerable sorrow about having to get a regular job.

A. H. Glenn and Associates in New Orleans, Louisiana, was a private meteorological and oceanographic consulting firm that served the offshore petroleum industry. They provided forecasting and oceanographic studies for any location in the world. They needed help, and I was the person. With employment waiting in New Orleans, my wife and I loaded our entire world into a small trailer and took off for the Deep South.

Working at A. H. Glenn was a bit like being back in the navy. Glenn maintained rigid procedures and methods. There was little room for innovation. But as is so often the case, things of great value frequently appear in disguise, only to reveal their true worth later on. At the time, it was inconceivable to me that four years with A. H. Glenn would in any way be of the slightest benefit, except for a monthly paycheck.

As it turned out, Glenn gave me four years of priceless education. I look back now and realize that I was working beside a genius, one who had mixed the complex sciences of meteorology and oceanography into a viable business. For his unique services, corporations and governments around the world paid large sums of money.

From a small office at the Lake Front Airport (on the south shore of Lake Pontchartrain in New Orleans), we forecasted the weather and oceanographic conditions for countless locations around the globe—from the Gulf Coast to the South China Sea to New Zealand to the North Sea to the frozen area of Spitsbergen to the ever-changing and violent Gulf of St. Lawrence and the Gulf of Alaska. A high percentage of our forecasts were used by offshore oil rigs and platforms, where the lives of oil workers were on the line. Glenn demanded the best you could give, and it had to be excellent.

He methodically taught me the techniques of evaluating and analyzing raw and sometimes limited oceanographic data and converting it into detailed, professional studies. We dealt with such things as ocean bottom pressure anomalies, one-hundred-year storm winds and waves, wave forces, storm surges, currents, and many other elements. I was assigned to complete oceanographic studies for locations all around the world.

As Glenn came to trust me, he assigned tougher projects. One nightmare of a project was for the giant Ekofisk oil field in the North Sea between England and Europe. Included in the oceanographic design information were the one hundred year storm wave heights and associated forces. The projected height of such a wave was a little over sixty feet from trough to crest and was expected to occur at least once every one hundred years. Unfortunately, after the steel for construction had been ordered, a wave of nearly ninety feet was recorded in the North Sea.

When the news of the unexpected wave became known, I received a conference telephone call from irate Ekofisk personnel at several locations around the world—Japan, England, and the United States. Everyone was yelling and most of them were swearing, some in English and some in other languages. When the commotion lessened and they allowed me to speak,

I gave them Glenn's phone number in Switzerland. That didn't work. It was the old bird-in-the-hand situation, and I was the bird. As the day faded into a very dark night and I grew even more weary of the verbal assault, the client's finally agreed that designing for a one-hundred-year storm was not wise, we all decided that they should redesign for a much higher-year storm.

It was crisis management experience that would later be invaluable to me when facing the wild weather antics of the southern plains. Before going to work for A. H. Glenn, I would never have been able to deal successfully with such a serious interpersonal challenge as I faced that day. I had been fortunate, though, to observe Glenn handle very significant problems during and after hurricane Camille.

Camille, in 1969, was one of the strongest hurricanes ever to strike the United States. We watched day after day as a tiny circulation bloomed into a raging fury as it moved from the southeast Gulf of Mexico toward the Louisiana-Mississippi coastline. Glenn, a person of few words (except when he was angry), commented early on, "This storm is going to be one of the worst, and it looks like its going to hit eastern Louisiana and Mississippi." Until almost too late, the official projections and warnings were for the Panhandle of Florida.

The Thursday before Camille struck, Glenn, after a long, intense session studying the latest surface and upper air presentations, announced very sternly, "She's headed right for the mouth of the Mississippi River." A chill rippled through my body like waves moving over the ocean. Writing quickly on his standard hurricane warning form, Glenn scribbled out the details. "Advise hurricane precautions and evacuation for all offshore clients from Louisiana to Mississippi by 6:00 P.M. Friday," Glenn commanded. It was a voice and look that I had never before seen from him.

Throughout Friday and Saturday the office was the scene of frantic activity. Handling the incredible volume of incoming calls was a lesson in panic control. The hurricane was now a powerhouse. Upper-air conditions indicated that it might grow to extreme proportions. Helicopters and crew boats shuttled hundreds of workers to shore, knowing that hesitation would result in tragedy.

By Saturday it was obvious to most everyone that an extreme storm with an abnormally high, deadly potential was churning through the Gulf of Mexico. Investigating aircraft reported continuing central pressure falls and associated wind increases. This giant low-pressure area was a killer in waiting.

On Saturday at 4:00 P.M., official hurricane warnings were canceled for Florida and extended west to include Louisiana. Just before midnight, the winds reached 100 MPH at the mouth of the Mississippi River in southeast Louisiana. Fortunately, evacuation of the residents of coastal Louisiana had already taken place by 6:00 P.M. Friday, the time Glenn had advised for our offshore clients.

Throughout Saturday night the storm continued its oscillating track toward southeastern Louisiana. Early Sunday morning, while fierce winds and titanic waves were raking southeastern Louisiana and hundreds of offshore facilities, Camille turned a little more toward the north. It appeared that the town of Pass Christian, Mississippi, on the southern coast of that state would bear the brunt of the chaos that carries the evil twins of the hurricane eye, wind and water.

Daybreak on Sunday morning felt like something out of a movie. With clear blue skies overhead and to the north, I could see on the southeast horizon a large, black, rotating disk that ominously grew larger by the moment.

Lake Pontchartrain was a few hundred yards to our north, and New Orleans, ten feet below sea level, was just out our front door to the south. I felt a bit insecure. As Sunday progressed, the force of the wind increased. By Sunday afternoon the eye of Camille was to our southeast. The wind screamed out of the northeast like a wounded beast, threatening to dump Lake Pontchartrain over the levies and into New Orleans.

The final glimmers of light on what was to be the last Sunday for over two hundred people in the path of the eye revealed a graphic scene. I stood on the south side of the huge, solid building that housed our office. Black, angry clouds ripped across the sky, spiraling inward toward a raging hell. The sound of the wind had become a steady, deep roar, interrupted only by the higher pitched scream of pieces of tin slicing through the air. Small rocks were lifted from the ground and propelled through the air with such velocity they shattered car windows.

As darkness and even stronger winds arrived, many tragic scenarios invaded my thoughts. Most hurricanes wobble along their path. Each time Camille wobbled to the left, I thought it was headed directly for New Orleans. I also thought about the storm making a sharp left turn once it moved inland. Glenn had told me that a hurricane moving west, north of Lake Pontchartrain, would totally flood New Orleans. Only my responsibility of plotting aircraft reports of the deadly track of the storm kept me from running through the building in fear and abandon.

Late Sunday night, as the storm took dead aim at Pass Christian, one of the aircraft reported "eye wall diameter fifteen miles, low-level flight winds: 175 knots." The smaller the eye of a hurricane, the stronger the winds. Camille had a tiny eye and winds close to 200 MPH! It was now like a giant tornado. There was little hope left for people who had decided to ride out the storm.

"Oh, my God," Glenn mumbled when he read the aircraft message. "There's not going to be anything left over there." His comments didn't help my state of mind in the least, but with all of the clients taken care of, Glenn calmly retrieved a well-hidden bottle of vodka, drank a couple of shots, and lay down on a cot. The winds buffeting our building were sustained at 90 MPH, the gusts were 120 MPH—and the man went to sleep!

Throughout the night and early morning hours I watched the movement of the storm. I was unaware of the horror taking place just ninety miles to our east. As Camille moved into shallow waters, small islands disappeared. As she approached landfall, the storm surge built to thirty feet. Half of one small town vanished into the boiling waters. An apartment complex in which several people who were having a hurricane party disintegrated. Only the foundation remained. Two men who tied themselves to a railroad overpass twenty-five feet above ground level drowned. Families that took shelter on the second floor of their stately, old homes were swept away in the furious onslaught.

Monday morning arrived with light blue, sunny skies, pleasant temperatures, and moderate breezes from the southwest. A telephone call from a client who had just completed an aerial survey of offshore properties arrived on the heels of that superb morning. "It's gone! The whole thing has disappeared," shouted the terrified voice.

He was referring to what had been the world's deepest-water offshore oil platform. It had been positioned in three hundred feet of water in the South Pass area, just off the southeast coast of Louisiana. Designed to house dozens of oil wells and scores of workers, it was a city of steel on the ocean. Now it was gone, and A. H. Glenn and Associates had provided the meteorological and oceanographic civil engineering design data. The client was very unhappy, but Glenn never acknowledged an alleged error unless he could be proven wrong by factual information.

Even though I had not yet arrived in New Orleans when the civil engineering data were provided to the client, I still had a personal interest in the loss of that magnificent platform. The previous winter I had provided the wind and wave forecasting for the two-week platform tow, by barge, from safe harbor to its rough-water destination. With millions of dollars riding on each forecast, I developed a rather close attachment to that load of cold, hard steel the length of a football field. It assumed an identity of its own. I watched over it like a newborn baby. It couldn't walk or talk and I, for sure, didn't want it to fall into the water.

As the platform silently rested on the bottom of the ocean, the battle of blame surged back and forth in a sea of shouts and threats. Glenn held his ground like a lion protecting an injured cub from a herd of jackals. Brilliant and tough, he was a formidable opponent for anyone who challenged him. Eventually, after weeks of detailed study and analysis, the verdict was reached. Glenn was not at fault. The oil company admitted that their original analysis of the sea bottom soil structure was faulty. The rocking motion of the platform caused by the huge waves and extreme winds of hurricane Camille caused soil failure, a result of hurricane-induced, severe bottom-pressure anomalies. Glenn looked at me and said, "Remember, it's never over until it's over." On this occasion and many others, his advice and training would prove to be some of the most valuable I ever received.

Even though the work in New Orleans was fascinating, television meteorology was still my dream. Several times while living in the Big Easy I applied for on-air jobs at various locations around the country, but no one would take a chance on me. No experience in television, no job.

A short time after Camille, I wrote to Alec Gifford, the highly regarded news director at WVUE-TV in New Orleans. WVUE-TV had no weatherperson or meteorologist on staff. It appeared to me to be a great opportunity. In reply, he wrote, "The reason we do not have a weatherman is that I question the need for this service." In an area raked by hurricanes and under what seemed like constant assault from flooding, he did not believe he needed someone to advise and warn his audience. Later, as I had more experience with news directors, I became very familiar with this attitude.

Throughout my time in New Orleans, the tornadoes of the southern plains beckoned. On April 30, 1970 at 1:00 A.M., a twister four hundred yards wide ripped through Mustang, Oklahoma, and then continued through the western and northwestern part of Oklahoma City. Forty-six people were injured, and fifteen hundred homes and businesses were damaged or destroyed. Just thirty minutes after the first tornado set down, another blasted its way through the Camelot addition near Lake Hefner and destroyed ten homes. There had been no advance warning. I believed I could make a difference if I were there.

Chapter 4

Out of
the Valley

Finally, in 1971, I got the courage to quit my job and return to Oklahoma. With stars in my eyes and belief in my heart, we rolled north from the swamps of Louisiana past the green, rolling hills of east Texas to the plains of Oklahoma. Our beautiful three-year-old daughter, Molly, rode with my wife and me. She called us Mister Gary and Miss Mary, loved beans and rice, and had never seen snow. Years later, my wife confided that she silently cried most of the way home. I was leaving a great opportunity behind and facing a future that had yet to take shape, except in my mind.

Like a beautiful spring being cut short by the intense heat of summer, the exhilaration of being home again was brief. The reality of not having a job and being refused every place I applied caused doubts about my abilities and fear about the future, feelings that were new to me. After a few months, I finally found a job selling advertising for a small circulation magazine on a commission basis. It was not only a bad idea, it was a real-life nightmare, night and day.

Fortunately, one day I noticed construction of a radar antenna at KTOK Radio, where my wife worked. Their place of business was a house on West Main Sreet in Oklahoma City. The radar antenna was the most beautiful sight I had seen in a long time. I was certain they would need me.

I made an appointment with the KTOK general manager, Hewel Jones. Excitedly, I told Mr. Jones that he would without a doubt be in need of my services as soon as installation of the radar was completed. Well, he informed me that my services would not be required—ever. Their disk jockeys were going to be the readers of radar.

A bit more than dejected, I went outside and climbed into my old Volkswagen bug with its bent frame and hole in the floorboard. I drove aimlessly around town for several hours. Numbed by the bitter cold air that flowed from the hole in the floor, I felt like a failure. My goal and dream since the seventh grade had been to be a television meteorologist. The reality, however, on that gloomy day was that I was nearly broke. I thought of my wife and child to whom I had promised so much but was delivering so little. Somewhere along the way during that painful afternoon of drifting, self-examination and self-pity, I reached a solid decision. Mr. Jones and KTOK Radio would need me!

As spring approached, I went every morning to KTOK and sat in the lobby, drinking their coffee because we had none at home. After a period of seeing me every time he turned around, Jones began to acknowledge my presence. I was there because I knew he would need me. I had to have a job. Eventually spring arrived and, along with it, severe weather.

Thunderstorms developed quickly one morning at 7:00 A.M. Large hail and high winds were reported in the western part of Oklahoma City. No one at the radio station knew how to evaluate the radar returns. Hearing the sound of heavy footsteps moving rapidly on the old wooden floors toward the lobby, along with the loud crack of thunder, I knew I had a job. Bursting through the door, Mr. Jones shouted, "Do you know how to run that radar back there?"

"Yes sir!" I replied.

"You're hired!" he yelled. "Now get back there and tell our audience what is happening with this storm!"

KTOK Radio was the first station in Oklahoma to own radar. It had a flat plate antenna with a five-degree beam width, originally designed for aircraft use. Identifying small weather features was impossible. The system had a very short range, and no storm examination and evaluation capabilities. But it was radar, and I had a media job in meteorology.

As the storms blew through Oklahoma City that morning, I made very nervous weather broadcasts from a tiny, cramped makeshift room with no ventilation and no chair. My shortness of breath and shaky voice undoubtedly frightened more than it informed the audience.

Just as the storms moved east of Oklahoma City, a tornado watch was issued for eastern Oklahoma. "Here's the county breakdown. Read it when I give you the cue," the disk jockey shouted as he threw the Teletype paper toward me. As I gazed at the counties to be read, I deeply regretted that I had never bothered learning to pronounce all of those county names properly. With a cotton-dry mouth and a pounding heart, I began to speak. During the next 120 agonizing seconds, I so totally garbled such names as Pushmataha, Okfuskee, Okmulgee, and Pontotoc that no one could understand a word I was saying.

As I continued, it was impossible not to notice the crowd of employees gathering to watch the new guy go down in flames. Their hysterical laughter and antics didn't help matters at all. My debut in radio was, to say the least, not promising.

On arriving for my second day on the job, I tried to slip unnoticed into the radio station. It didn't work. They noticed, and they all started laughing again. For a fleeting moment, I thought maybe I was in the wrong business. Finally, Jones stepped forward and separated himself from the horde that was having such fun at my expense. "Follow me," he instructed.

I followed him to what had been the garage of the home that housed the radio station. He reached up, pulled on a cord, and like magic, stairs appeared. "This," Jones said, while pointing to the dark rectangular hole in the ceiling, "is your office and radar room." It was the attic.

My office was in a small, dusty environment with a partial floor, a desk, telephone, microphone, and a lone light bulb dangling from a time-worn wire attached somewhere in the dark recesses above me. The radar display was propped up on the desk with a two-by-four piece of lumber. On first glance, it was pretty gloomy. It didn't take me long, however, to realize that an ugly office is much, much better than no office at all.

In the business of radio and television, the public must know your name and like you or you'll soon be back in the category of no office at all. In an effort to raise my name recognition from the depths of total obscurity, Bob Riggins, the top-rated morning drive-time disk jockey, suggested that I be called the Wonderful Weather Wizard. That move helped some, but my broadcasts were stiff, boring, and uninteresting.

In the narrow halls lacing the old house that was the radio station, management personnel suddenly developed the ability to pass by me as if I were a ghost, invisible to the eyes of others. It was obvious that something had to change or I was not going to last long in the radio business. Sensing that unemployment was on my horizon, Riggins again came to the rescue. "Gary," he said, staring at the floor, "when the weather is not serious, you must learn to laugh and have a good time on the radio." With that he led me, like a lost puppy, into one of the control rooms and turned on the recording equipment.

"Laugh," Riggins commanded. My face was made of stone. "Laugh!" he shouted. I uttered a weak, forced noise that sounded like a pervert making an unwanted phone call. "Laugh, darn you!" Riggins screamed. He startled me with the last yell so much that I jumped backward and nearly fell out the door and into the hall. At that point, we were both laughing. "That sound you just made," he said, "was a genuine laugh. Remember what it sounds like and practice an hour a day until it becomes natural to do it on the air."

Thanks to Riggins I kept my job that spring, but summer, and with it fewer storms, was rapidly approaching. Soon I was called into the general manager's office and informed that I would not be needed when spring ended. I considered this a very bad sign. Then out of the broadcast control room and down the narrow halls came Riggins to the rescue.

"Gary, to survive, you need some help," he counseled in a quiet, serious manner. "I've seen that lizard you caught at Canton Lake—and, by the way, it isn't a good idea to go around scaring all of the secretaries with that thing." I acknowledged this with a timid nod of my head. Continuing, Riggins said, "I think we should call your little green lizard friend an 805-pound thunder lizard and I further believe you should integrate him into your broadcasts when the weather is not serious." Thus was born a half-fictitious, half-real radio character that carried me through the summer.

My lizard partner made frequent early morning, creative vehicle chases on the interstates. The audience loved it. Many people stopped and called from various locations, reporting tongue in cheek, that they too had been victims of one of the imaginary chases. The thunder lizard changed colors with the weather and was so ugly that frozen turkeys in the local grocery store ran from him. My popularity grew rapidly, and I still had a job—until the end of the summer.

The general manager then informed me, "Thunder lizard or not, you definitely will not be needed during the fall." I remembered Glenn's words, though—it's not over until it's over. Quickly I began gathering statistics. Determining the number of high school and college football games played each week along with the number of persons attending was vital to my case. Also critical was my finding out how important weather information was to each and every one of those people attending. Armed with solid information and a written presentation, I nervously met with the general manager to convince him again that he did need me. I kept my job.

In late September 1972, while sitting in my attic office, the phone rang. "Weather Wizard Central," I answered.

"Gary, my name is Jack DeLier and I run KWTV here in Oklahoma City," the voice on the phone said. At this point, my excitement bordered on panic. The voice continued, "Our weather guy, David Grant, is leaving. I've heard you on the radio, and you do sound a little wild and crazy, but we would like to talk to you."

"Yes sir!" I responded, in my best navy tradition. We quickly made an appointment, and concluded the phone call, and I then fell on the floor. It was a dirty floor, but I didn't care. Maybe, just maybe I thought, my lifetime dream was about to come true.

Finally the day of the appointment at KWTV arrived. To say I had jitters in no way describes how I felt. I met with a man named Jack Sallaska, director of programming. He was a personable, soft-spoken individual who gave me one of the best pieces of information I ever received about television. "Television weather," Sallaska said, "is like walking along the top of a narrow fence. On one side is total boring meteorology and on the other is total entertainment. Fall off either side and you're out of the game. You will not succeed. The key," he continued, "is to balance yourself on the fence and carefully mix the two sides depending on the weather situation." We then set an audition date and time.

Terrified at the thought of the audition, I arrived at 7:00 P.M. on the designated Tuesday night in a mental state that one might describe as near hysteria. A slow walk through the television station to the broadcast studio with Jack Sallaska helped me settle down.

As I stared at the weather set, I could see it consisted of two army green metal displays, each four feet by three feet. The two areas housed large, rotating, four-sided metal drums. On the drums, a person could place flexible, magnetized strips for cold and warm fronts and any variety of numbers and letters. White chalk was used for on-air notations. The forecast was presented with magnetized letters and numbers. By today's standards, it was nearly prehistoric. At the time, it was the rage.

Prior to doing three takes on the audition, Sallaska loosened me up with his comments and facial expressions. It was a nervous half hour, but a fun one. "That's a wrap," Sallaska informed me. "It's three rolls of the dice in this business and you're either in or out," he added. With that comment, I became aware for the first time that the connection between the present and the future is but a thin string.

Back in my attic office the following week, the phone rang. "Gary, this is Sallaska. The boss would like to meet and talk about you coming to work at KWTV." Only the end of the world could have kept me from that meeting. Suppressing my excitement, I agreed to the meeting in less than five seconds.

I drove up to KWTV at 8:00 A.M. on the following Friday morning. My old Volkswagen bug backfired, shuddered violently, and then shut itself down. Some of the employees who were walking into the building ran, others ducked, and others just stared in amazement.

As I haltingly entered Jack DeLier's office, I noticed that it was larger than my apartment. With a warm but serious demeanor, Sallaska introduced me to DeLier and the news director. Both DeLier, the general manager, and Sallaska were pleasant and friendly as we gathered around a huge, wooden desk. DeLier, a classy and gracious person with sixty-eight combat missions over German territory during World War II, was also very direct and forceful. I liked him immediately.

The news director, for whom I would work, sat slumped in his chair. It didn't take long for him to make it clear he considered meteorologists unnecessary. DeLier firmly reminded everyone that he was in charge, a welcome relief to me.

"Gary," said DeLier, "I've seen your audition and I like it. We want you to come to work for us." I could scarcely believe what I was hearing.

"Well, Jack, I'll need some time to think about it," I replied.

Ignoring my response, DeLier continued, "What kind of money would it take to get you to be our chief meteorologist?" My heart rate must have been setting a world record at that point.

In my hands was a chance at my lifetime dream, yet I continued to make strange responses. "Mr. DeLier, if I should take this job, and I'm not sure I will, I'd have to have a thousand dollars a month. And besides, I really like radio."

DeLier stared at me. He made no change in expression for what seemed like ten minutes. I couldn't move a muscle. He was like a big cat in the jungle locking in on its prey. "Gary, you go home and give it some thought and call me next week," DeLier very pointedly instructed me. "If it's a yes," he continued, "we have a deal."

In a few moments, I was back in the real world, with two helpful persons shoving me and the Volkswagen bug down the driveway. With a loud bang the engine caught. The car lurched forward to about 15 MPH, sounding more like 80. Passing by the front of KWTV, I could see people looking out the windows. My little car roared off, leaving behind a thick plume of exhaust.

My wife was excited. Our combined income did not equal a thousand dollars a month. Yet I was hesitant. "You know Mary, I really like radio because you can be creative and have fun on the air." That was true but a bit unexpected as I had wanted to be on television since the seventh grade. As the evening wore on, I thought of every possible reason not to take the job at KWTV. All of my comments were being generated by my fear of actually being on television. In the end, my wife and common sense prevailed. I took the job.

Chapter 5

Real
Television

On my first day at KWTV, I carefully looked at the weather office for the first time. It was in studio A, a huge cavern of a room from which the news, weather, and sports were broadcast. Lights hanging from the ceiling grid system gave the appearance of a multitude of upside down floor lamps. Covering the concrete floor was a maze of large cables strewn about in what might have been a random distribution.

The weather office, though, was not really an office. It consisted of a battered, lonely looking desk with a cracked glass top. The desk was shoved up against a thirty-three-pound tile block studio wall, forty feet high and fifty feet wide. On that bare north wall, stuck on unevenly spaced nails, were a few weather maps and several oddly tilted columns of Teletype paper.

The broadcast set and newsroom were one and the same, located in and around the south wall of studio A. When I inquired as to why the weather warning Teletype was against the south wall, some fifty feet away from the weather office, I was informed that the news director thought it looked good there. When I asked why the weather map machine was not in the studio but, rather, down a hall and up some stairs in a small space behind the studio A control room, I was told that it made too much noise to be in the broadcast studio. "Not smart," I silently mused, because during severe weather, quick access to data is crucial.

Little did I know that I had walked into a world and time where television news directors were considered independent and powerful. They had the largest number of employees in the company, the largest budgets, and egos to match their responsibilities.

News directors were positioned as protectors of the people's right to know. Their accountability was taken seriously. No one ignored a news director's order. Not being a submissive person, I knew there would be rough waters ahead.

That first day was full of surprises. "I'd like you to meet our new news director," DeLier said. "Yesterday he was our assistant news director." One day on the job and I was already on my second boss!

It was obvious that the news director was talented and bright. It was, however, also obvious he was not much interested in television weather.

The operational hierarchy of television, with respect to weather, was general manager, news director, assistant news director, managing editor, executive producer, show producer, and finally the on-air talent. It was not surprising that many instructions did not resemble the original orders by the time the information reached the designated recipient.

The year was 1972 and the Vietnam war was still taking its horrible toll not only on the troops but on the very fabric that had held this country together. KWTV at that time was no different from any other business organization. It was a mixture of military veterans, hippies, war protesters, and just regular folks who had no opinion about the war or at least didn't make one known. With such diverse convictions, our bitter differences frequently boiled to the surface, often disguised as other problems.

"Okay Gary," the news director very authoritatively stated, "I want you to practice for a week, and then I'll put you on the morning show. When I think you're ready, I'll probably let you do the prime-time evening shows." My pulse quickened, and my face tingled. I had been hired to be chief meteorologist to do the evening prime-time shows, not the morning shows. However, I did not want to lose my job before it started, so I kept my cool and forced myself to nod acknowledgment of his instructions.

Since the departure of David Grant, Lola Hall had been doing the prime-time shows. Bright, blonde, and very capable, Hall was a very popular television personality. She was to show me around and help me get started on my practice sessions. She was very gracious and explained everything very well.

After practicing every day for a week, I proclaimed myself ready for television. DeLier said it was fine with him and notified the news director that I would be doing the morning and noon shows for a while. The news director let it be known immediately that he was the one who would decide what I did and when. He let me know the chain of command in no uncertain terms and that I was never, ever to go around him again. Beads of sweat popped up across the news director's forehead as he continued to berate me. "The next time you do something like this," he screamed, "you'll be out on the street and I'll see to it you never have another job in television!" I stood motionless and said nothing. On the outside I was calm, but inside I was becoming profoundly angry. I knew if his furious attack lasted much longer, I would reach my flash point. I felt my adrenaline building. My response to fight-or-flight situations had historically been to fight. Exhausted from his tirade, he abruptly told me to leave his office. Quickly and gladly I obliged.

"Standby!" The camera operator shouted. With those words the thought flashed through my mind, "I really don't want to do this." Without hesitation, the ex-hippie behind the camera pointed his finger at me and the small red light attached to the front of the camera lit up. My mouth opened, but nothing came out. I'm sure I resembled a fish out of water.

It was my first live broadcast, on the 7:00 A.M. farm show. Everyone in the studio was staring at me. They were apparently waiting for some sort of noise just to confirm that I was still alive. Finally the word "fantastic" gushed forth. In fact it came forth several times before I began to make any sense at all. When that three-minute show was over, I was shaking from head to toe and felt weak as a newborn kitten.

The early morning farm show terror continued for a week. During the second week, it all became more natural as I became more relaxed. Still, it would be a long time before I could do a broadcast without becoming nervous before, during, and after it. During my first two weeks on the air, the news director was no place to be found.

On the second Friday I informed Jack DeLier and the producers that I was ready and would be doing the 6:00 and 10:00 prime-time shows starting the following Monday. Hall did the morning and noon shows on that Monday. I came in all polished up with a brand new maroon jacket, ready for the shows at 6:00 and 10:00 P.M. The news director was nothing short of ballistic—and rightly so, as I hadn't obtained his permission to start broadcasting on the prime-time, Monday-through-Friday shows. Despite his initial graphic objections, he did allow me to start the prime-time shows that Monday night.

The metal weather maps on the large four-sided drums somehow looked larger that night. Each drum weighed 180 pounds but felt much heavier. Every time I turned a drum, some of the letters and numbers would fall off or would assume a crazy tilt and have to be rearranged. It was frustrating, but those days, the norm.

The news format within which we worked seemed old-fashioned, drab, and boring. News anchors never smiled. They droned on in a monotone about stories they and some very inexperienced person had decided were interesting. It was not news that was interesting to the audience.

Audience rating services confirmed the lack of popular interest in our product. It was a terrible hour of news, running from 6:00 to 7:00 P.M. The objective, it seemed, was to make the viewers feel worse at the end of the show than they had when it began. Then, the entire hour was shortened and repeated from 10:00 to 10:30 P.M.

My weather shows were in stark contrast to the news presentation. With no radars or computers and a storm of chalk dust floating in the air, I smiled, laughed, stuttered, and stumbled my way through each show. I believed, and still do, that if you can deliver the information and also cause someone to smile, you have accomplished something good.

My style was revolutionary for the time and caused a considerable stir. I frequently talked of my 805-pound thunder lizard. My coanchors could scarcely believe anyone would talk about such things on television. The audience response was incredible. Hundreds of letters and phone calls came in, nearly all positive. I was considered good, but I didn't know it, or really care. I just wanted to have fun while doing my job.

Within a short time, the pressure from the public to view my thunder lizard increased to a point that something had to be done. I decided on a contest, asking the audience to draw, paint, or build an image of what they believed the lizard looked like. To help the project along, I interviewed one of our camera operators on a cold, windy November day concerning his claim that he had seen the famous creature.

The man was a classic. His long flowing hair whipped in the wind from under a big cowboy hat firmly pulled down to his ears. The collar of his too-small, sheepskin coat was turned up and snugly secured, covering his neck and cheeks. His tummy hung over a colorful Navaho belt that barely held up his baggy, paint splattered britches. On his feet he wore old, scuffed-up, steel-toed cowboy boots. He looked like a man lost in time, a combination of hippie and cowboy.

"Sir, I understand you saw something strange recently," I stated.

Without hesitation, he responded "Yeah man, wow, like just last week I was rollin' home at about 2:00 A.M. Sunday morning and this creature, you know, like ran across the road in front of me. I was in my car," he continued, "but you know it was so big that I've been driving my truck ever since!"

"Well, sir," I inquired, "just how big do you think it was?"

For a few moments he stared at the ground and finally said, "I figure, like, you know, man, it weighed about 805 pounds."

It took all I had to keep from smiling. Then I asked the big question. "Sir, just what do you think this creature was?"

With the harsh north wind buffeting us, he stared at the ground, kicked several rocks from the gravel-laden driveway, and then looked directly into the camera. "Mr. England, I think it was, without a doubt, an 805-pound thunder lizard."

The interview ran on the 6:00 P.M. weather. A disclaimer stated that the person about to be seen "thought" he had seen something strange. Audience response was quick and far beyond our expectations. Telephone calls jammed the circuits in our exchange for over an hour. I learned a valuable lesson that day. Some people who are watching the news and weather may be only half listening to what is being said.

Many of the callers thought a large wild creature really was running loose. Others called to say the interview was funny and great. Five people called to say that they, too, had seen the thunder lizard—and they weren't kidding!

The next day the commotion continued. Some calls were similar to those of the night before, but many were from people at their places of employment. They had not seen the show but had been told about it. That meant new viewers for KWTV. Even representatives from a UFO organization called. Somehow my fictitious creature struck a warm and friendly cord with the television audience of Oklahoma. I had stumbled onto a promotion manager's dream.

The station received an avalanche of contest entries. There were thousands of pencil drawings, water color and oil paintings, charcoal renderings, and sculptures made of plaster, wood, and metal. Over 80 percent of the entries were from adults. Many were absolutely terrific and were undoubtedly created by very talented people.

We awarded the prizes for the best entries and quietly put the thunder lizard forever to bed, I thought. Surprisingly, questions about him have continued to the present day.

As I was new in the television business, I did not think much about the tremendous response to the contest. The news director was livid. I hadn't told him what I was planning. The general manager, though, was pleased, as was the promotion manager, who quickly tried to lay claim to the idea. The "newsies," as I called them, treated me as if I had a contagious disease. They kept at a safe distance and barely acknowledged my presence. Most of them stared at the floor as they passed me in the hall. I think they were appalled that a I was bringing humor, fun, and happiness to television news.

The news director constantly confronted me. He thought I was, day by day, destroying the credibility of his news programs. His argument was weakened, however, by the fact the latest audience ratings revealed that thirty thousand more homes were now watching KWTV news, weather, and sports than had been before.

Soon a new news anchor appeared on the scene. Ralph Combes was bright, articulate, and witty. Women thought he was very handsome; I just thought he was tall. Ralph and I hit it off well, launching the era of so-called "happy talk" television in Oklahoma.

Combes constantly, in jest, made caustic but very humorous comments about me. He did a lot of "short" jokes, referring to my height. The audience responded in growing numbers. They loved the crazy exchanges between Ralph and me. At the beginning of the news, we sat side by side. During one particularly funny moment, I laughed and at the same time reached over and placed my hand on his. He slowly turned from me, looked directly into the camera, and said, "Mr. England, you're touching me."

The other male news coanchor was a fine person, but was definitely from the old school of news. I always thought that somewhere along the line, aliens had extracted every ounce of humor from his body. He did not approve of anything Ralph and I did.

Ted Leitner was the sportscaster. Big in stature, very knowledgeable, and possessing a great sense of humor, he would eventually go on to be a top television sportscaster in southern California.

Bob Jenni was our resident naturalist. He took great pleasure in bringing live and sometimes dangerous animals into the studio and onto his news segment. The word "panic" could not adequately describe the reaction of everyone in the studio when three of his rattlesnakes slithered off the anchor desk and onto the floor. Jenni calmly finished his segment with one remaining snake still in his grasp. He then coolly hunted down the others as the news program continued.

When Jenni was going to talk about snakes, which was fairly often, he kept them in a sack just inside the back door, the door we all came through. It was also an area that was dark. Even when I knew there were going to be snakes in a sack just inside the door, it still scared the wits out of me when the room suddenly filled with the sound of angry rattlesnakes.

At this point in my career, I was always happy just to get through my weather shows without some major disruption. I did not always accomplish that goal. One evening, on the air, I turned from the weather board to the camera to finish my show and make the toss back to the news anchor and found myself face to face with a llama. It had big, dark eyes and bad breath. Or course, standing beside it with a smile the size of a half-moon was Bob Jenni.

It was such a surprise that I could do nothing but laugh. Then Ralph Combes began to laugh. I walked out of view of the cameras, leaving Combes with the llama. Combes was a city boy, and the story he had to read next was about limousin cattle. As far as he knew, those were cows in a car. The llama moved closer to him. Every time he tried to read the story he would start laughing, tears streaming down his face. Then, abruptly, the llama relieved itself, right on the six o'clock news.

An avalanche of calls hit the station switchboard. A majority of the callers loved the humor. The news director, finding no joy in what he had witnessed, exploded in anger.

The following day on the six o'clock news, something happened on our television station that had never happened before. Sitting between Combes and me was the news director. He begged the audience to forgive __WTV for what had happened the previous night. Then he asked the audience to call or write the station and vote, yes or no, on whether Combes and I should be allowed to continue our antics.

The response was overwhelmingly positive for Combes and me. The news director had been, in fact, very crafty. While reinforcing the seriousness and credibility of KWTV news, he had let the audience see him as the bad guy. He created sympathy and support for Combes and me, resulting in higher ratings. That had been his goal, and he attained it.

On the few occasions during those first months when I had to issue a weather warning, I would run from the weather office across the studio to the Teletype, rip off the warning, and run back to the weather set. Totally out of breath, I would come on the air, panting, and issue the warning. My lack of composure frightened a lot of people, so I had the telephone company move the Teletype line to the weather office, adjacent to the weather set. The news director was not pleased when he discovered the change, but he made only minor rumblings.

In early 1973, as my first television spring storm season approached, I proposed to Jack DeLier that KWTV purchase weather radar. I soon met with him and John Griffin, the station owner, to discuss the matter.

Griffin was a fascinating man. Warm and friendly, he was a person who cared about people. A busy man, he nevertheless took time to support his employees. Not long before our meeting he had sent me a memo stating, "I notice you are getting a lot of fan mail on your weather. Here's another. I think you are doing a splendid job." He didn't have to send anyone such a memo or note, but he did.

The news director was also well known for his frequent memos. His memos, though, rarely contained praise. His messages usually bespoke a high degree of frustration, likely due to the difficulty of converting concept to successful product.

This memo from the news director can serve as an example:

"I just watched the first three minutes of our news program and decided it was probably time for a few pointed words."

For the past three weeks or so I have had the incredible feeling that this staff has contracted some exotic disease resulting from the placement of one's head in a most unlikely spot, resulting in severe pain and agony.

I think the word that characterizes our daily effort too often is pedestrian. The first definitions I find in Webster's for that word is unimaginative, commonplace. And that about sums it up.

I'm not talking about those occasional lapses that hit everybody. And I'm not talking about those "crunch" times when our backs are to the wall. I am referring to the rather common occurrence on our news programs of errors that result from negligence, apathy, indifference or some other sign of the same malaise.

So, let's make a deal.

Monday, January 29, 1973, will be a fresh new day starting a fresh new week. Let's all pretend that we didn't see the stupid display that passed for a news program on Thursday, January 25. Beginning that day I will absolutely not tolerate such a display ever again.

I expect to see that damned pig pen of a newsroom cleared of old boxes, assorted junk and the collected scrap of weeks.

Also I expect to see some professionalism and some attention to detail.

Although I've got a lot of fish still to fry, you can expect to see me out in the newsroom more.

To my knowledge, John Griffin, never saw any of the news director's memos. He was always too busy ensuring that KWTV was on the leading edge of technology. Therefore, it took very little effort on my part to obtain his approval for the radar purchase. The only question was which radar to buy.

One radar was the type I had used at KTOK Radio; the other, a superior Enterprise Electronics system. The Enterprise system had a price tag of fifty thousand dollars. Griffin was a little uneasy about committing to such an expensive radar when the other cost just twenty thousand dollars.

I urged Neil Braswell, the creator of Enterprise Electronics and its radar system, to come to Oklahoma City and meet with Griffin. Braswell was a brilliant engineer and a pretty fair salesman, too. He agreed, and we all soon met. As the meeting wore on, the blunt-spoken Braswell finally said, "Mr. Griffin, if you don't have the money, we can work out a payment plan."

"Write him a check!" Griffin snorted. And with that, a revolution in television weather began.

Chapter 6

First
Exposure

On February 26, 1973, KWTV installed the newest Enterprise radar, the predecessor to the radar system the U.S. Weather Bureau would acquire the following year to fill in the gaps in its radar coverage.

I could barely contain my excitement as the unprotected, domeless radar antenna was installed on the roof directly above the newly constructed radar weather office. The office was in a corner and, at twenty-four square feet, was very tiny. As I sat in the darkened office, my hands shook when we powered up the system. I could scarcely believe it—radar images right before my eyes!

Within a short time, a powerful winter storm raked Oklahoma with fierce winds and snow. "Look at this! Look at this!" I shouted over and over, sometimes to no one but myself. It was a riveting experience actually to see weather in motion. Suddenly the antenna quit rotating, and about the same time, someone in the studio shouted, "Hey look! Snow!" Yes, snow was coming right into the studio.

At the crack of dawn the next morning we swarmed over the roof of the building. There was our brand new antenna, tilted at an odd angle. Six feet in diameter, it had taken the full brunt of the wind and toppled over, ripping open the roof to which it was bolted.

The general manager erupted with the force of a sizable volcano. The antenna had been bolted to a tin roof with layers of tar on top. "Only a fool would do this," he yelled. "Get it fixed and get it done now!" he added. His voice echoed off the adjoining buildings. "And get your rear in there and order a dome to cover this antenna, too!" he further instructed the unfortunate person who was standing closest to him.

As the spring of 1973 arrived, so did severe weather. In March, Oklahoma experienced seven tornadoes and what seemed like hordes of severe thunderstorms that packed high winds and hail. In April, the frequency and intensity of the storms increased. As usual, many of the official severe storm warnings were issued after the fact. It seemed as if the only way someone a mile down the road could be warned was after some other person upwind had already been hit. The warnings were simply too late—sometimes issued so late that they should have been classified as news stories. The people issuing the warnings were good, but the technology was not. Even the acquisition of real-time severe weather information from the field was erratic at best.

As one of the people delivering the warnings, I received a considerable amount of negative feedback. All too often, I heard or read the comment "You always come on after the storm has already passed." It was mild compared to some of the criticism. Unfortunately, it was a true.

I soon decided that I was not going to take the blame for the tardiness and inaccuracy of the official warnings. If I were going to receive blame or credit, I would earn it on my own. That decision would result in a long-lasting rift between me and the U.S. Weather Bureau.

With the help of several ham radio operators, I launched a major effort to organize and train a large number of volunteer storm spotters. The information they provided, along with a high volume of input from the civil defense, police, and fire departments, resulted in vastly improved information from the field. It saved many lives. KWTV reporters, photographers, and engineers were also trained in storm spotting and safety. When I suspected tornado potential, the trained observers were positioned close to the expected storm path. The storm chasers proved their value on May 24, 1973.

The day dawned warm and humid. Conditions were reasonably favorable for severe weather, but not ideal. During the afternoon a line of thunderstorms developed along a cool front. The line was advancing through northwestern Oklahoma toward the southeast. Because of the potential for tornadoes, I alerted the newsroom of the situation. We sent one KWTV crew northwest, toward the line of storms, and another to the west where, as of that time, no thunderstorm activity had developed. A few ham radio spotters were also informed of the situation. Outside the television station, we could see large white towers of clouds gracefully billowing toward the heavens. At 1:30 P.M. a tornado watch was issued by the National Severe Storms Forecast Center (of Kansas City, Missouri) for a large part of Oklahoma.

A tornado watch is issued when atmospheric conditions suggest the possibility of tornadoes. It usually covers several thousand square miles and is in effect for about six hours. A tornado warning is issued when a tornado is sighted or indicated on radar. A warning usually covers one or two counties and is in effect for an hour or less.

The Teletype soon began pouring forth official tornado warnings for locations along the line of advancing thunderstorms, but no actual tornadoes had been sighted. As the storms pushed toward central Oklahoma, every few minutes I would switch from the 125-mile range to the 50-mile radar range. It was a standard action to detect any storms developing close to the metro area. On one of those standard reviews, I noted a new radar return about 40 miles to the west-southwest of KWTV.

The new storm was growing rapidly, and it was in advance of the main line of thunderstorms. Frequently, the storm that develops away from and in front of the main line of thunderstorms is the storm that may produce a tornado. With each increase in size of the storm, we received more information via television station and ham radios as well as the civil defense, police, and fire department radio scanner. I was learning very quickly that one of the keys to success during severe weather was being able to listen to multiple sources of information, make decisions, and issue instructions, all the while keeping my composure.

The radar continued to probe the growing storm. Thousands of times per second the transmitter would send a burst of tiny, invisible waves of energy, 5.5 centimeters in length, and listen for any return of that energy. Any returned energy reflected back from the rain and hail was processed by the radar receiver and displayed on the scope at various levels of intensity directly related to the amount of energy received. From those displayed levels of intensity, it was obvious that the storm was continuing to gain strength rapidly as it moved from west to east.

"Five-five-eight to car two," I said on the station's radio channel.

"Car two, go ahead, Gary," came the quick reply.

"Car two, a ham radio operator reports a lowering. It's a possible wall cloud west of Union City. Can you confirm?"

After what seemed like an eternity, the radio came to life again. "I see the lowering, something hanging down toward the ground, but I can't tell what it is for sure. But it does not look friendly," was the reply.

A wall cloud is a lowering of the cloud mass, usually in a rain-free area in the right rear quadrant of a thunderstorm. If it is rotating, most likely it is an extension of the rotation inside the thunderstorm and can produce a tornado.

"Five-five-eight to car two, can you confirm rotation?"

"Negative, five-five-eight. From my vantage point, I am unable to confirm rotation, but this is a nasty-looking storm."

As I talked with our unit in the field, ham radio spotters confirmed rotation. In order to evaluate the thunderstorm, I flipped a switch to remove some of the lighter precipitation from the radar screen, a technique known as attenuation. I was looking for an area in the mid-levels of the storm that had no radar returns or an area with light radar reflectively surrounded by heavier precipitation. I increased the attenuation factor even more. There it was, a donut-shaped presentation at twenty thousand feet, a bounded weak-echo region. This was a radar feature associated with large-scale rotation in the updraft region of the storm.

I lowered the antenna to monitor the storm at around ten thousand feet. With minimal attenuation locked in, the appendage protruding from the back southwest quadrant of the storm set off an alarm in my head. It wasn't a classic hook echo, but with the mid-level bounded weak-echo region and associated spotter reports, it was enough. Without hesitation, I shouted into the intercom to the control room that housed the director and audio person. "Tornado warning! Go now!"

In a voice totally lacking in stability, I broadcast the warning: "Radar here at channel nine indicates a tornado may be developing just to the west-northwest of you folks in Union City and Minco. This storm is moving toward the east. All residents in the path of this possible tornado should take immediate tornado precautions. If you don't have a cellar or basement, go to the center part of your house, lowest level, smallest room, preferably a closet or bathroom. Cover up with blankets and pillows! Stay with channel nine, we'll keep you advised."

I grabbed the station radio microphone. "Five-five-eight to car two, go for the storm! Go for the storm!"

"Care two to five-five-eight, I'm on Highway 37 moving toward the storm. It looks like it's going to do it!"

At about 3:30 P.M. spotters reported a well-developed wall cloud with dust swirls on the ground beneath it. You don't have to see a funnel to have a tornado—this was evidence of a tornado on the ground! Within minutes, the twister was causing damage. The extremely dangerous storm turned slightly to the right and continued to grow in strength and width. It was taking dead aim at the small town of Union City.

Once again the KWTV warning beeps sounded. "You folks in Union City, if you haven't taken shelter, do it now!" I shouted. "Our spotters," my quavering voice continued, "report a large tornado very close to the northwest section of Union City. It's moving toward the southeast. As I said before, take cover now!"

Just northwest of Union City, the devastating swath of winds spread to nearly a quarter mile in width. Just before 4:00 P.M., the twister's path width narrowed to about six hundred feet as it entered Union City. With the smaller diameter, winds increased to an estimated 250 to 300 MPH along the center of the tornado track. Devastation along the center track was complete, but many lives were saved because it was daylight and there was considerable advance warning. There were, however, two fatalities and four injuries.

After my second warning broadcast, I realized that no official warning had been issued before I went on the air with the first warning. My reaction had been an automatic response to a dangerous situation, a result of military training and those intense years with A. H. Glenn—experience I had undervalued at the time.

The Union City tornado slithered its way east and soon dissipated. No sooner had the storm passed when rumblings of discontent from official quarters reached KWTV. The message was clear. I was out-of-line for issuing a tornado warning before the official warning was issued. It seemed to me that by reacting quickly and decisively to a deadly threat, we had saved lives. I learned an easy lesson that day. When on the cutting edge, expect an avalanche of criticism. If I were going to enjoy the glory, I would also have to endure some personal and professional attacks.

The next day, John Griffin made it all worthwhile. In a memo to the staff, he wrote,

> I thought our live coverage of the tornadic activity yesterday was the best I have ever seen. It certainly should prove to our viewers that we are the most thoughtful television station regarding the safety of everyone in our coverage area. Also, getting the crew to Union City as fast as we did and getting coverage on the air denotes to me an awareness of our obligation to our viewers. I thought the coverage was superb.

Griffin's memo became the framework for my future beliefs and actions with respect to severe weather and television: those of us in broadcasting have an obligation to provide the audience with fast, accurate severe weather coverage, with the safety of one person being as important as the safety of many.

As important as the Union City tornado was to my career, it was also of great value to the National Severe Storms Laboratory (NSSL) in Norman, Oklahoma. For the first time in history, a thunderstorm that produced a major tornado was probed and recorded by experimental Doppler radar. The storm was also recorded by conventional radar, and its life cycle photographed from birth to death. The Naval Electronics Laboratory Center, located in San Diego, California, also made valuable lightning measurements. Documentation of the storm was thorough, and the detailed research that followed was nothing short of revolutionary.

Study of the rather primitive Doppler data gathered by the NSSL during the Union City tornado confirmed that within a thunderstorm producing a large tornado, there is a recognizable vortex pattern. Such a large-scale vortex structure, or rotation, after meeting several Doppler radar requirements, is designated a mesocyclone. Scientists discovered that wrapped within the mesocyclone are smaller, even more intense rotations. Those tightly wound winds represent the actual tornado, called the tornado vortex signature (TVS).

The mesocyclone is an elongated vertical tube of spinning low pressure inside the thunderstorm, usually located between the updraft and downdraft regions of the thunderstorm. As the circulation lengthens and narrows, the wind velocities around it increase. Visible evidence of a mesocyclone can sometimes be observed in the form of a rotating wall cloud. A wall cloud is a lowering of the cloud mass, usually in a rain free area on the right-rear flank of the thunderstorm. In a major tornadic thunderstorm, the mesocyclone rotation extends from the wall cloud up to the top of the storm. A mesocyclone will produce a tornado less than 50 percent of the time but approximately 90 percent of the time will produce some type of damage, ranging from tornadoes to large hail to high-speed straight-line winds.

Doppler radar, like conventional radar, sends out short wave bursts of energy thousands of times per second. Between each burst, the radar receiver listens for any reflected return of the energy waves that are bouncing off of precipitation particles. Doppler radar, in addition, then analyzes received energy for any shift in frequency with respect to the transmitted energy waves. From those shifts in frequency, the speed of movement of the particles can be determined.

As one stands near a highway listening to the sound of a vehicle passing, the vehicle sounds different as it approaches than it does as it moves away. This change is known as the Doppler effect. It occurs because the sound waves produced by an approaching object are compressed into a higher wave frequency (producing a higher pitch), while those of a receding object are lengthened, producing a lower wave frequency (and lower pitch).

With Doppler radar, the speed of the precipitation is determined by the difference in the frequency shift between the moving particles. The speed of the precipitation parallel to the radar beam is proportional to the extent of the measured frequency shift. Precipitation that crosses the Doppler beam at a ninety-degree angle is invisible to the radar. Therefore, the only complete wind measurement on Doppler radar occurs when the precipitation particles are moving exactly parallel to the radar beam.

The application of Doppler radar to severe weather was in its infancy in 1973 and would for many years progress quietly in the realm of pure scientific research. Few people outside the research community were aware of the existence of Doppler radar, and it is likely that even fewer dreamed of the incredible positive impact it might have on short-term weather warnings. I know of no one besides myself who dreamed of having Doppler radar at a television station.

Throughout 1973, the fertile tornado grounds of Oklahoma gave birth to seventy-five tornadoes. Thirteen fatalities occurred along with over two hundred injuries. Late in the storm season, my decision to broadcast not only the official warnings but also our own warnings had brought about a small improvement in the quality of severe weather warnings. Extended advance warnings, though, such as the one for the Union City tornado, were rare. Surprise tornadoes and unexpected straight line wind damage continued. Also, it seemed that every time a thunderstorm appeared on the horizon, a false warning would be forthcoming. The nature of the business invited attention and comment.

Chapter 7

Outbreak

Working on the forefront of television weather means being the focal point of criticism. It is part of the job to handle the many good-natured comments directed to the station about no warnings or too many warnings. It is also part of the job to be polite when receiving negative responses from dissatisfied viewers.

The day after a large and deadly tornado in Texas my phone rang. "Why didn't you warn them?" came the monotone question.

"Well," I said, "those people and that town are in Texas. They are not in our broadcast area, which means they do not receive our television signal."

"Why didn't you warn them?" he asked again. Keeping my temper under control, I managed to get off of the phone.

Moments later the phone rang again. "It's me again," he said. "Why didn't you warn them?" Having just experienced two, long stress-filled days of severe weather, I was already near my limit.

"I already explained that, so here's what I'll do. We receive some good calls, and I log those as positive calls. We also receive some calls like yours, and I log those as negative calls."

At that point there was a long silence from my strange caller. Finally he said, "But I don't want to be a negative call."

Feeling very frustrated, I pleaded, "Just leave me alone, please." He never called back.

A few viewers seem to extract considerable joy from verbal attacks on television meteorologists. In the early years, newsroom personnel seemed to enjoy such endeavors, too. Those television journalists and photographers who blessed the weather department with their personal views on television weather surely were not born cynical. Perhaps living and working in an environment that deals with death, destruction, and human suffering left them in need of an outlet for stress and frustration.

My colleagues covered a war whose vicious appetite for the young consumed nearly sixty thousand Americans. They interviewed politicians who could not or would not answer questions. They produced stories on people who were cheats and thieves, sons who in the silence of the night stole from their fathers. They reported on criminals who served a scant portion of their sentences and were set free, only to prey again on the public. Their job was to condense bad events and evil people into the lifeblood of the newsroom.

Part of the Vietnam war era was the increasing usage of illegal drugs. Television was not immune. While watching live television in the early to mid-1970s, one might see, on occasion, the studio floor or wall. Camera shots were sometimes shaky and out of focus. Long shots of nothingness rather than pictures were common. On a fairly regular basis, the video transmitted did not match the audio. During periods between news shows, it was common to see in our parking lot illegal smoke pouring out of a vehicle window. It was a bizarre world.

The first and only time I saw a real drug dealer he was standing in the control room for studio A. His long, purple trench coat hung from his shoulders to the floor. A wide-brimmed, flat-topped matching hat covered his head. He was a friendly looking man with a huge smile and pearly white teeth. Gathered around the fancily dressed dude were several of my coworkers.

Always curious, I walked into the room and said, "What do you all have in here?" The smiling man turned toward me. The crowd around him parted.

"Anything a person could want," came his husky reply. When my eyes moved from his smiling face to the open box he was holding, I was shocked. In the container was an assortment of hundreds of multicolor pills. "What do you need?" he asked. Stunned, I mumbled something about having to go to the bathroom and quickly exited the room.

The drug problem was present in the television industry as it was all across the country, but slowly and carefully KWTV management obtained evidence and identified the persons who were using drugs on the job. It was a long process, but eventually the problem was eliminated at my work place.

Against the backdrop of drugs, humor, and a war, the world of television continued to move. With his quick wit and sharp tongue, anchor Ralph Combes continued to hammer me on the air. Ted Leitner continued superb sports coverage. The news director continued to fret over the smallest of details. And our audience ratings continued their upward spiral.

Successful severe weather warnings did not spiral upward. Failure to warn in advance was more the rule than the exception. Late in the evening on November 19, 1973, a large complex of thunderstorms moved through the central Oklahoma skies. A deadly tornado cloaked itself in the darkness of night and the roar of thunder. Although it was unidentifiable as a potential killer on conventional radar, I issued a severe thunderstorm warning.

A severe thunderstorm is generally one that has hail three-quarters of an inch in diameter or larger and winds of 58 MPH or greater. Unknown to me and everyone else in the business, this thunderstorm had secretly given birth to rotating winds approaching 200 MPH.

The twister instantly snuffed out two lives at Blanchard, located roughly thirty-five miles south of the KWTV radar room. With blinding speed, the winds blasted northward. Near the end of its twenty-four-mile life, the tornado ripped through parts of Moore and then into the southern edge of Oklahoma City. Beneath the rubble in Moore were two more lifeless bodies. In Oklahoma City another life was claimed.

Frantic calls for help filled the airways. The sound of fire trucks, police units, and scores of ambulances echoed across the landscape. As the search for the dead and injured continued at an agonizing pace, information slowly filtered into the newsroom. Five dead, fifty-three injured, and scores homeless.

It made no difference that the official warning had also originally been for a severe thunderstorm. I knew that some of the people who suffered that night had put their trust in me, and I had not been able to help them.

It was a very difficult period in my life. My emotions ran out of control. Powerful feelings about myself shifted from moment to moment and from day to day. In my mind, graphic pictures of the entire scenario came to life. Beckoned forth by some uncontrollable force, the deadly movie ran through my thoughts like a raging river. I doubted my ability and that of the radar. I questioned why some had a future while others died. The memory of that terrible fall night so many years ago is always with me, each and every storm season.

Good humor, fortunately, is an effective healer. At KWTV there was absolutely no lack of humor. Some of the humor was distasteful, but most was spontaneous, great fun.

When one of the news anchors was having a birthday that required a large number of candles on the cake, problems arose. We lit the candle-covered cake and prepared to present it on the 6:00 P.M. news show. Within moments, the weather maps caught fire. Smoke poured from the weather office. The news anchor continued to read his stories, while we desperately ripped the flaming maps from the wall, threw them on the floor, and stomped on them until the fire was out. Then we turned our attention back to the cake. In just a few moments I was to present it to the news anchor.

I slipped into my chair on the news set to wait. At the appropriate time, I gently slid the cake toward the anchor. "Happy birthday," I said. Before he could respond, the entire cake caught on fire. It was incredible. Live on the air, a small birthday cake was flaming like a raging inferno. The news director also lit up, but not with happiness.

That news director, my second boss, soon departed under pressure from top management, a victim of a self-inflicted wounds. He had ordered one too many investigative reports that top management did not agree with. I never saw him again. Soon my third news director was appointed.

On April 3 and 4, 1974, during a period of twenty-four hours, 148 tornadoes occurred. They blasted across Illinois, Indiana, Michigan, Ohio, southwest New York, Kentucky, West Virginia, Virginia, Tennessee, Mississippi, Alabama, Georgia, western North Carolina, and far northwest South Carolina.

In this, the largest recorded tornado outbreak in history, 315 persons were killed and over 5,000 were injured. An abnormally large number of the twisters were judged to have winds of over 158 MPH. Thirty-four storms were classified severe, with winds of 158 to 206 MPH. Twenty-four earned the classification of devastating, with winds ranging from 207 to 260 MPH. Six rare "maxi-tornadoes," with winds 261 MPH to more than 300 MPH, were also identified.

As I watched and listened to the drama unfold from the security of Oklahoma City, I was astonished at the magnitude of the disaster. It was an unparalleled display of raw, savage power. The area covered was so huge, it was not only unexpected, but not believed possible. The sheer number of tornadoes alone seemed to make the event an impossibility that actually happened.

In the world of television, it seemed that unlikely events and impossible actions took place every day. One photographer on his own, quietly and unnoticed, recorded illegal activities. When his presence was finally noticed, he shouted, "See you in jail!" With that comment, a wild chase erupted on the streets of Oklahoma City. Eventually, the employee arrived at the television station, a shaken and temporarily repentant person. The excitement of that day though, would soon pale in comparison to the events of June 8, 1974.

The early morning sky on June 8, 1974, was crystal clear but heavily laden with moisture from the Gulf of Mexico. By midmorning, the beautiful green trees of spring leaned to the north under the force of increasing southerly winds. Noontime brought temperatures in the eighties. Small wisps of low stratus clouds flashed across the sky, mindless passengers on winds of 50 MPH.

As the sun heated the ground and the ground heated the atmosphere, shimmering plumes of clear air quickly became hotter and less dense than the surrounding air. The hot air sprang upward. The hot spots, miniature low pressure areas, immediately began to rotate.

Shortly past noon, the atmosphere could no longer contain the mix of volatile ingredients. Cumulus clouds silently appeared in the blue sky. The storms were forming just to the east of a dry line, a north-south zone with very hot and dry winds pressing eastward against the energy-packed moist air covering Oklahoma. The rising air, caught in cyclonically rotating updrafts, cooled and condensed into tiny pieces of moisture, taking the form of puffy-white columns thousands of feet tall. A phase change from water vapor to tiny pieces of liquid water released heat. The latent heat of condensation overcame the natural cooling process of rising air and allowed the lifting towers to continue to be warmer than the surrounding atmosphere. Remaining buoyant, the twisting of the upward-bound air intensified. The rising air encountered vertical winds increasing in speed and changing in direction from the southeast at ground level to the southwest in the higher levels. That vertical wind speed and wind directional shear quickly gave life to strong rotating updrafts that would soon evolve into full-blown meso-cyclones and then into tornadoes.

"Tornado on the ground! South of Mustang, just southwest of Will Rogers World Airport!" said the nearly panicked voice over the ham radio.

"Hit the buzzer! Hit the buzzer!" I screamed to one of our meteorologists. His large hand smashed against the box, activating the intercom that connected the weather room to the control room.

"Yes," came the director's bored reply. He had been quietly positioned in his chair for over two hours.

"Priority one, tornado warning, take pointer over radar!" I yelled. "Priority one" signifies that the warning is an emergency. The television program is to be interrupted without hesitation, regardless of what is being broadcast.

"Standby, priority one, tornado warning, take pointer over radar. Are you ready Gary?"

"I am ready."

The "Are you ready?" and "I am ready" verbiage delayed operations for a second or two but was vital to clear communication. "Go" and "no" can sound a lot alike. A few years before, one of our meteorologists and the director confused the two words. The meteorologist was punched up live on the air while shouting, "No, don't go!" From that point on, we developed very rigid and precise procedures for broadcasting weather warnings.

Awaiting the director's final commands, I held the lever in my hand that controlled the pointer used to show the audience exactly where the storm was located.

"We're in the beeps, Gary. Cut it and go." The pulsing tone of beeps abruptly alerted the audience that an emergency weather bulletin broadcast was imminent.

I rattled off the warning information. "A tornado has been reported just southwest of Oklahoma City near Will Rogers Airport. All residents should take immediate tornado precautions. If you don't have a cellar or basement, go to the lowest level, smallest room, center part of your house. Get down on the floor and cover yourself with blankets and pillows!" It was 1:42 P.M.

Over the ham radio receiver came the next report: "Tornado has just hit the Will Rogers weather office and is moving toward the northeast into Oklahoma City." In a matter of moments, ham radio reports, muffled scanner messages, phone calls from the public, police transmissions, and radio information from our own units were creating a deafening, nerve-jangling din within the walls of our tiny radar room.

"Gary," someone notified me, "the Weather Bureau has been knocked out and Jim Williams at WKY says he's taking over the warning responsibilities."

Jim Williams, who had been named chief meteorologist at WKY-TV after Bob Thomas left, was a highly respected and fine meteorologist. I held him in high regard. I was, however, surprised that he was issuing warnings on his own.

For nine miles the tornado consumed and then spewed out the remains of homes and businesses in southwest Oklahoma City. Just as I realized it wasn't the dreaded big one, severe thunderstorms mushroomed all over the area.

At 2:11 P.M., another strong tornado slammed into Spencer, a sparsely populated suburb of Oklahoma City, just to the east of the television station. The entire area was now under a tornado warning. "Car three to five-five-eight. We have another one on the ground just west of Jones. There is considerable debris in the air. It's at least a half mile wide!" a frightened voice reported over the station radio. It was now 2:18 P.M. Jones, another small suburb of Oklahoma City, was just to the northeast of our location.

I loved weather, but at this point I was wondering why I had gotten into this business. Momentarily, a brief calm slipped over the area, and then suddenly the lull ended. The second wave hit Oklahoma like a hammer. At 3:48 P.M., a small but vicious tornado spun to the ground just southwest of Choctaw. This location, just to the southeast of the television station, was also a suburb of Oklahoma City. The expression, "Too close for comfort" came to mind.

"Large tornado on the ground just southwest of Drumright!" The unidentified radio transmission split through the intense noise of the radar room. I shuddered. The deadly message from the field confirmed what I was looking at on radar. A large, classic hook echo seemed to cover most of the screen. It was 3:55 P.M.

"Priority one, tornado warning update!" I screamed into the battered intercom. I had issued the warning for Drumright earlier. Now with confirmation from spotters, I made one last frantic effort to warn of the impending disaster.

The Drumright warning sirens screamed out in the face of the approaching messenger of death. With winds approaching 250 MPH, the 400-yard-wide aberration of nature quickly silenced the sirens and fourteen lives. Born from the morning's innocent blue skies and destined to return to nothingness, this tornado, the strongest of the day, was to stay on the ground for forty-five miles and take two more lives.

With numerous severe thunderstorms and possible tornadoes in progress in our viewing area, the television screen was alive with warning after warning. From a stiff, hard chair, I evaluated the radar returns. From that same stiff, hard chair, I had been delivering warnings and updates for over two hours. My eyes burned, my back hurt, and I could feel my heart rapidly beating in my chest.

In quick succession from 4:46 P.M. to 5:45 P.M., five more twisters slammed down in our broadcast area, just to the east of Oklahoma County. The first was 1,000 yards wide with winds near 200 MPH. The last of the five ripped a path 450 yards wide and twelve miles in length, with winds near 200 MPH.

At 5:50 P.M., two tornadoes moved through Tulsa. With those tornadoes over ninety miles northeast of Oklahoma City and not in our viewing area, we allowed ourselves to hope that the storms were about over for our highly stressed crew.

Then, testing us to the limit, another tornado snaked to the ground just north of Seminole and cut a twenty-mile path across the countryside. As the Seminole twister moved northeast, a thunderstorm approaching the southeast part of the Oklahoma City metro area abruptly intensified.

"Gary! Viewer reports a funnel cloud just southwest of McLoud!"

"Priority one, tornado warning, take pointer over radar!"

"Are you ready?"

"I am ready."

"We're in the beeps, Gary. Cut it and go."

With a dry, thick tongue I blurted out the warning. At 6:50 P.M. the tornado touched down just south of McLoud. After terrorizing humans and animals for two miles, the twister quietly evaporated.

By 7:00 P.M., about six hours after the siege had started, there was no longer a threat to our viewing area. As I rose from my chair, my back muscles cramped, and rivers of pain shot along my spine. Forcing myself to stand up straight was very uncomfortable. My pain, however, was nothing compared to the pain of those whose lives were scarred by the tornadoes.

Later that evening two more tornadoes touched down in far eastern Oklahoma, bringing the total number to sixteen. There were 294 people injured, 15 were killed. It had been an exhausting, disastrous day. Again I wondered why I was in the television weather business.

Chapter 8

Promoting the Product

Working in television weather is more than storms and warnings. It encompasses a broad spectrum of exciting experiences and activities. The quiet months allow time for recovery from the rampages of severe weather and for those other, nonemergency activities.

Once of the more important of these is promotion, the marketing of television personalities to the public. Such efforts include television promotional announcements, fund raisers, speeches, parades, and many other functions. Some public appearances can be as wild as the spring storms.

"It's for you Gary. Some guy at a radio station says he has to talk to you," said a meteorology intern, handing me the telephone.

"This is Gary. How may I help you?" At the other end of the phone, a man from a radio station made a most unusual offer. His words and mine are burned into my memory.

"Say, Gary, the Purcell rodeo is coming up, and we would like for you to be involved."

"Just what would you like me to do?"

"Ride a bull."

"No, I do not care to ride a bull."

"Wait a minute, Gary. We have a trained bull."

"How does that work?" I asked, and in so doing made a very serious mistake.

"You put the bull in the chute, open the gate, and he will walk out. Then you blow a whistle, and he will buck, and then, on blowing the whistle again, he will stop bucking."

I thought about it for a few seconds and said, "You have a deal. I'll have the whistle in my mouth."

Believing my riding a bull at a rodeo to be a splendid promotional opportunity, I took advantage of every opening on radio and television to talk about the upcoming event. Of course, I never mentioned that the bull was trained.

About a week before the rodeo, a package arrived in the mail. Inside the parcel was a book on bull riding. Fanning briefly through the pages, I soon tossed it aside. My bull was trained—I needed no such instructions. I did notice one sentence: "Bull riding is best described as violent, random motion."

Finally the day of the rodeo was at hand. I wore a large white cowboy hat. Tucked into my bell bottom pants was a bright red, white, and blue vertical striped-shirt. I didn't own real cowboy boots, so I wore glistening blue, patent leather ones. I looked like a city dude.

An experienced, timeworn cowboy was assigned to get me ready. He spit his big chew of tobacco out and smiled a very sly smile. At that point, he didn't have to say it. They had tricked me. I was going to have to get on a wild, bucking bull.

"Let's get your spurs on," he instructed me. He smiled again. I noticed he had three teeth missing. I wondered if he had lost them in a bull ride. He also walked with a limp. I was certain he must have been gored by a bull. He strapped the spurs into place. My time had come.

In the chute was fifteen hundred pounds of bellowing, thrashing, ill-tempered flesh on the hoof. Bending down, I peaked between the boards for a better look at his face. I could see one huge bloodshot eye and one horn. He saw me and smashed up against the side of the chute. The bull knew he had a city dude and knew I was afraid of him.

Having mentioned this appearance many times, I had talked myself into a corner and couldn't back out. With great mental agony, I climbed up the side of the chute. From above, the bull looked huge. I noticed he had a horn on the right that matched the one I had seen on the left.

"Ease down on him, really careful like," my now-serious helper urged. The instant my bottom met the bull's back he made a frenzied attempt to hook my left leg. Frightened beyond description, I tried again. Same results. Finally my huge friend with the bad attitude allowed me to sit on his back.

He just stood there, no movement except for the heaving of his giant lungs. I cautiously slid my legs down his sides. He shifted just a little. I felt his powerful muscles ripple along my legs and buttocks. I knew, without question, this would be my first and final ride.

My helper wrapped a leather strap tightly around my right hand. Breathing in shallow gasps, I realized he was, in fact, tying me to the bull. As I became even more tense, I pressed my spurs against the tough hide of the creature. He didn't like it.

Within a split second, the bull exploded in a raging fury. He lowered his rump and slammed hard against the back of the chute. With my hand securely anchored, my torso and head were thrown back into a nearly horizontal position, and correspondingly my feet flew forward toward the bull's head. He had me prone on his back. In the next split second, he desperately tried to hook my left leg with his left horn. Unsuccessful in that effort, he leaped up and pitched forward in a well-practiced maneuver that reversed my position. With my hand as the pivot point, my upper body flew forward. My face smashed into the back of his neck. When I finally got myself back into a vertical position, the camera operator from KWTV, a Vietnam veteran wounded three times, looked at me and said, "Don't do it Gary! Don't do it!" But it was too late.

With one hand tied down and the other in the air, it must have been fear alone that caused my head to nod, similar to an uncontrollable cold shiver. The gate swung open. With a deep, loud bellow my powerful host leaped sideways out of the chute. The moment his hoofs hit the ground he spun left. Immediately he reversed his turn and threw his massive body high into the air. At this point, my only contact with the bull was my right hand. The rest of me was in the air. I then slammed into his back, face down. At the same moment, he dove forward, planted his front feet in the sod, and with a mighty heave, threw his rump and me up and forward.

In an eruption of dirt, I plowed into the soft sod directly on my head, rolled over, and ran for the fence. With my once-beautiful hat now a mashed remnant of its former glory, I climbed the fence and went home. I now understood the meaning of violent, random motion. I also had a better understanding of the limits I should maintain in promoting myself.

The following Monday morning, Jack DeLier spotted me shuffling down the hall. Every part of my body had been bent, stretched, or crunched. My slow, methodical movement was designed to mask my condition, But DeLier was not a casual observer.

"Gary, let's have a word or two," DeLier crisply stated.

"Yes sir," I mumbled. It even hurt to move my mouth.

Standing tall, with power and authority in his voice, he boomed out his instructions, "Don't ever do anything that crazy again!"

"I won't," I blurted out, mingling his last few words with mine.

"Your responsibility," DeLier said, "is to make public appearances that are somewhere within the range of normal behavior. Your chief responsibility is to provide severe weather warnings and to do that, you must stay healthy. You got it?"

"I've got it, Jack," I responded. Turning smartly to walk away, my battered left knee gave way. I staggered and fell against the wall. DeLier had made his point.

Responsible for the safety of untold numbers of people, I always had the possibility of violent weather on my mind. A powerful force, it over-shadows life outside of television. No matter the time of year, the threat of severe weather is a persistent, silent companion.

On a bleak, winter day in 1975, I analyzed the surface and upper atmospheric conditions. A cold front of moderate intensity was moving southward over Oklahoma. An upper-level storm system was to our west in New Mexico, moving toward the east. To our south, warm moist air on the surface was restricted to far southern Texas.

The forecast seemed obvious. It would turn sharply colder as the cold front shoved southward. The upper storm system would move over western Texas and Oklahoma. The combination of upward lift from the cold front and the upper system would initiate precipitation. A period of freezing drizzle or freezing rain would be followed by snow as the temperature of the air fell below freezing at all levels. It appeared to be a standard scenario for such a weather pattern.

All went according to schedule, or at least appeared to. By midnight the cold front had moved south of the frequently dry Red River that represents the border between northern Texas and southern Oklahoma. Temperatures in southern Oklahoma were in the upper forties, and dew points were in the upper thirties—too cold for tornadoes, which typically require temperatures in the seventies or higher and dew points in the sixties.

In the dark of night, unknown to anyone, the advancing upper storm system to our west created a swift, low-level jet stream. The low-level jet, a band of strong winds about two thousand feet above the surface, rapidly funneled the warm air in southern Texas northward. Energy-laden air, silently rising up over the cold surface air, was then lifted even more by the approaching dynamic system from the west.

Shortly after midnight, people in southwestern Oklahoma noticed furious lightning flashes. The phone rang in the forecast center.

"Forecast center, this is Gary. How may I help you?"

"England, I'm in Altus. What the hell is going on?" shouted the caller. "There's lightning all over the place, and the wind is terrible!"

It's not uncommon to have overrunning thunderstorms with cold air on the surface, resulting in thunder snow, but this was different. I could feel it. The sound of his voice and the noise of the storm raging in the background created a frightening moment.

"Sir," I responded, "on radar, there are potentially severe thunderstorms in your area."

"Power is off!" he yelled. "The winds are incredibly strong."

Locking the phone between my shoulder and cheek, I hit the buzzer. I gave instructions for a priority-one severe thunderstorm warning. I issued the warning, careful to mention that tornadoes come from severe thunderstorms.

"I just put the warning on," I told the caller.

"Thank God," he responded. Then in mid-sentence, the line went dead.

The entire southwestern part of Oklahoma was now lit up with severe thunderstorms. It was 12:30 A.M., and the first tornado was on the ground in Altus, killing two people and injuring twelve.

The storms roared toward the northeast with large hail and strong, damaging, straight-line winds. At 1:00 A.M., another person was killed just west of Snyder by a second tornado. By 3:00 A.M., two other tornadoes ripped narrow but damaging paths to just south of Oklahoma City.

It was extremely unusual. Temperatures were still in the forties. Snow was beginning to fall, and tornadoes were tearing through the cold, night sky. In meteorology, it is wise to expect the unexpected.

Early the next morning the news director dispatched several crews across the winter landscape. The damage they reported on amounted to nearly ten million dollars. The cost in human suffering was also high. Four individuals had been killed, and over 120 were injured. It was a scene usually reserved for springtime.

Chapter 9

Close Encounter

As the world of television continued to spin, research continued at the National Severe Storms Laboratory (NSSL) in Norman. Scientists worked methodically toward the creation of a tool that would locate and identify severe weather before it struck. Creating the first operational Doppler radar was a lofty vision; each storm that could be probed and analyzed brought the NSSL closer to the goal of continuous, real-time use of Doppler radar for severe weather prediction.

During the spring of 1975, the shutdown time required for the NSSL Doppler to switch from one wind-speed range to another was reduced from about thirty minutes to less than two minutes. As a large number of tornadoes have a very brief life span, this was a very significant step.

Just as Oklahoma weather is a magnet for those interested in severe storm research, it beckons relentlessly to a very unusual breed of individuals called storm chasers. Each spring, storm chasers pour into Oklahoma in search of a close encounter with a tornado. For some, the chase and the encounter become the one, all-consuming goal in their lives. For others, it only takes their vacation time each year. And for at least one storm chaser, it is a business. He transports tourists along the highways and dusty back roads in search of this extreme thrill.

On June 9, 1975, my phone rang. "Mr. England, my name is Jim Leonard. I'm a storm chaser from Florida and I've taken a picture of a tornado that I think you will want to see."

"Why is that?" I asked.

"It's spinning the wrong way," he responded.

He was correct—I did want to see the picture. I had been taught that tornadoes spin only cyclonically (counterclockwise) in the northern hemisphere.

Jim Leonard ambled through the door a few days later. A student from Florida, Leonard did not know at the time that the tornado mystique had him firmly in its grasp. A glimpse into his future would have shown him face to face with tornadoes and hurricanes for the next twenty years.

Leonard had taken still pictures and a Super-8 movie of an anticyclonic tornado near Alva, Oklahoma. In doing so he had, at least to my knowledge, become the first person ever to document and therefore prove that a tornado can spin clockwise in the northern hemisphere. It was quite an accomplishment.

I introduced the wrong-way tornado man to my news director. Leonard said a quick hello and disappeared out the door. Being confined indoors was not for him. In the vast, open plains, there were more roads to travel and more storms to find. Out there, at an unknown place and time, waited the ultimate tornado experience.

This news director (my fourth in three years) was always very serious about his responsibilities, but once, for a brief moment, I was able to loosen up his tightly wound world.

On Thursday, January 8, 1976, and for several days before, a heavy, frigid air mass lay over Oklahoma—air that had arisen and matured in the darkness of the vast arctic regions. The bitter cold penetrated everything it touched.

On my television shows, I cautioned everyone about the dangers of such cold air. I carefully detailed the precautions they should take for personal safety in their cars and homes. I felt like I had provided a fine service. I was very proud of myself—so proud that I forgot to follow my own instructions.

Wednesday morning, January 7, was not a pleasant one at my home. The water pipes were frozen solid. After many attempts, I finally located a company that could get to my house before spring. Bud at Bud's Plumbing and Heating Company said he could do the job, and he did.

Relieved to have running water again, I went to work. Our newscasts and weathercasts that day were focused on the bitter cold and all of the associated problems that came with it. Careful to tell no one about my personal crisis at home, I completed my 10:00 P.M. show and went home.

The next morning I arrived at work early to get a head start on another busy day. When I walked into the newsroom, I saw the news director with several employees gathered around.

"Hey, England," shouted one of the individuals, "have you seen this morning's newspaper?"

"Sure haven't," I replied. I looked at the news director. He smiled. At that moment, I knew someone had discovered my secret.

"Mrs. Harrington," the *Daily Oklahoman* stated, "reported that one of the bearers of the ill winter tidings, television station ace thermometer measurer Gary England, was among her firm's frozen pipe customers Wednesday." It took quite a while for the ribbing to cease, but at least my problem had injected some humor into an often stressful environment.

Television operations by nature include a considerable degree of conflict. It is inherent in the system, and normal in a world where differing opinions abound.

Most news directors I have known believe in short weather shows. Most television meteorologists, myself included, are convinced that longer weather shows are necessary. I remember so well the day I reminded the news director that audience research definitely concluded that people watch the news just to get the weather. He became very agitated. In fact, for a few moments, I thought I might actually witness the first human self-explosion in history.

Even more difficult than my day-to-day interaction with the news director was the annual budget process. Usually, the news director must approve any capital expenditure requests from the weather department. The budget process is like trying to take a bone from a hungry dog. If you get the bone out of the dog's mouth, it bites your leg. Due to the pain you drop the bone. The dog then retrieves it, and the cycle begins again. It was a process I disliked intently.

It was budget time, early 1976. I believed KWTV needed two new items to stay on the cutting edge: number one was a weather satellite machine, and number two was a powerful new color radar system. Weather satellite picture loops had made their way to television by this time, but they were black-and-white and of very poor quality. I had located a company that produced a device that would receive and produce high-resolution weather satellite photographs every fifteen minutes. Sample pictures revealed a resolution so fine that on a clear day, some of the highways in Oklahoma were visible. The second item on the wish list, the new color radar system from Enterprise Electronics, offered a larger antenna and much-improved severe storm evaluation capabilities.

My proposal did not impress the news director. Budget battle lines surged back and forth. The news director fiercely fought all attempts to allocate funds for items other than those he believed directly enhanced his operation. Ours was a long and heated discussion.

Fortunately, the final decision was made by John Griffin and Jack DeLier. Griffin read my proposal, looked at Jack DeLier, and said, "Do it." When the news director started to speak, DeLier quickly raised his hand and firmly announced, "This meeting is over."

In early 1977, just after the appointment of my fifth news director, one of my long-awaited deliveries arrived. The thrill of opening the huge packing crate brought back memories of Christmas mornings long past. The new satellite weather machine inside would represent a major advance in television weather. I christened it StarCom 9, mixed the chemicals, loaded the film, and plugged in the power. Fifteen minutes later, I was the proud recipient of a seven-by-eight inch, high-resolution glossy photograph, a picture that had been taken by a spinning weather satellite over twenty thousand miles above the earth. The news director saw the picture, but showed not the slightest interest.

The StarCom 9 photographs were instant hits with a television audience that was accustomed to seeing only blurred satellite images. They revealed, in intricate detail, small cumulus clouds as well as towering thunderstorms. One could even see the shadows of the clouds on the ground.

On February 23, 1977, strong westerly winds over western Texas and Oklahoma lifted tons of dry topsoil into the air and created a major dust storm. With StarCom 9 I could see where the dust was and easily determine where it was going. As the dirt swept over central Texas and toward Louisiana, I called Pat Shingleton, a television friend in Baton Rouge.

"Pat, this is Gary England. How would you like to look really good to your audience?" I asked.

"Talk to me," was his immediate response.

"Tell your audience there will be a dust storm in Baton Rouge tomorrow."

"Right. Like a truck can fly."

"Pat, listen to me. Tell your audience that the big duster is coming tomorrow."

"You're serious, aren't you?"

"You got it. Do it and you'll look so good you won't be able to stand yourself."

Shingleton crept out on that weather limb and forecast a dust storm. The dust came to Baton Rouge. To his audience, Shingleton seemed like a miracle worker, a prognosticator of great vision and talent.

The satellite pictures were fabulous. I felt compelled to keep the news director aware of our significant technological advance, so I frequently slipped one of the images under his office door after the 10:00 P.M. show.

Even though my encounters with now five news directors taught me that most had little time or affection for meteorologists, the coverage of severe weather was something on which we could agree. After a severe weather event, there were no tough decisions. Damage had to be reported, and it was easy to locate. Sadly, the need for weather coverage on May 20, 1977, was obvious to everyone.

With KWTV's new, color radar system installed and operational, tornadoes began in southwestern Oklahoma in the early afternoon. Soon buildings, homes, and airplanes were splattered across the land. Altus Air Force Base was first to be hit. The storms then boiled northeastward toward Oklahoma City.

By 5:00 P.M., angry-looking clouds covered the horizon. Lightning danced through the sky. The distant rumble of thunder grew louder. Radar, focused on a storm to the southwest of the station, revealed the beginnings of cyclonic circulation. Several tornadoes had already touched down, and it appeared that this storm would join the day's total, which would finally be sixteen.

In the dimly lit radar room, I watched the drama of tornado generation unfold. "Bounded weak-echo region, overhang, maximum top over the weak-echo region, low-level hook echo," I whispered to myself.

"We have a supercell here. Standby for a priority-one tornado warning," I cautioned the director.

"Standing by for priority-one tornado warning," replied the voice on the other end of the intercom.

The storm was positioned a little west of due south from the television station, near Blanchard, Oklahoma. In search of prey, it moved toward the northeast. I couldn't wait any longer.

"Priority one, tornado warning, super upper third, take pointer over radar. I am ready," I barked into the intercom.

"Priority one, tornado warning, super upper third, take pointer over radar. Are you ready?" came the nearly rote reply.

"I am ready." I quickly snapped.

"We're in the beeps, Gary. Cut it and go," shouted the director.

The huge thunderstorm moved over parts of Norman. A frantic caller informed me that his lawn chairs had been lifted skyward and were currently rotating in the air to his northeast.

The raging storm turned to the left. Its track was now due north. Big hail producers usually turn left while tornadic thunderstorms are more likely to turn right. This one had turned left and was moving along Sooner Road, just to the southeast of the television station. Even though it was a left-turner, tornado activity appeared imminent. Within minutes the storm was well into Oklahoma County, spreading over sections of Midwest City and Del City.

The station radio came to life. "Car one to five-five-eight."

"Go ahead, one," I said.

"I'm in position as instructed, just north of Interstate 40 on Sooner Road. I'm in heavy rain. Also have fairly large hail."

His comments sliced through the noise of the radar room and hit me like a hammer. "He's in trouble," flashed through my mind.

"Five-five-eight weather to one!" I shouted into the microphone. "If you're near the large hail, you may be close to a tornado!"

"This is car one. I can't see a thing. The winds are increasing, rain is torrential, and the hail is scaring the crap out of me!" shouted the photographer in car one. He was not from tornado country and was having the experience of a lifetime.

Again we heard the frightened voice over the static-filled station radio, "This is one! This is one! Rain is decreasing. It's still hailing. The wind is awful. I can now see to the south."

Without a doubt, he was directly in the path of a potential tornado. "Five-five-eight weather to one," I said in a voice that was quivering as much as the voice from car one. I released the transmit button. In that fraction of a second, a desperate transmission from the field stopped all of us in our tracks.

"Oh my God! Tornado! Tornado!" The radio fell silent.

His message shook me to the bone. Somehow I managed to keep my cool and broadcast a tornado warning update. While I was delivering the new information, the phones were ringing, people were shouting instructions, and someone kept trying to raise the now-quiet car one.

At that moment, we later found out, the photographer in car one could see the tornado in the air coming directly toward him. Fortunately, it passed over head well above the ground. He jumped from his car, camera turned on, and firmly in hand. The ferocious winds instantly slammed him to the ground. He was not going to be defeated. Lodging himself between the door and the frame of the car, he pointed his camera north. In his viewfinder, a gas station fifty yards away was torn from the face of the earth. Within seconds the tornado became wrapped in rain and was no longer visible.

Once back at the television station, the jubilant photographer proudly displayed his just-developed film. He had managed to shoot ten seconds of usable footage. As we viewed it for the first time, everyone was surprised to see he had captured a multiple-vortex tornado. On the ground directly in front of him, three tornadoes had been tearing through the gas station. He was indeed fortunate to be alive.

Chapter 10

Changes in the Wind

During the spring of 1977, Oklahoma tornado frequency was above normal. Fifty-four twisters were recorded. We were busy. Organized chaos was the norm. Extremely long hours at the radar, panicky phone calls from the public, the difficulty of directing tornado chase crews, and the sheer noise level in the radar room took a physical and mental toll on all of us.

In Norman, thirty miles to our south, the NSSL was conducting the first operational warning experiment involving Doppler radar. The experiment was designed to test the advance warning time of the NSSL Doppler radar against the warning lead time provided by the National Weather Service (NWS, formerly the U.S. Weather Bureau) in Oklahoma City, which used conventional radar.

Fascinating results caught my attention immediately. The average lead time during the test for NWS tornado warnings was one minute. The NSSL advance warning time using experimental Doppler radar was twenty-three minutes. Twenty-three minutes might allow those in the path of a tornado to reach shelter, an opportunity that did not exist with the current warning system. Years of methodical and difficult work by many dedicated persons at the NSSL facility had produced findings that would eventually revolutionize severe weather warnings to the public.

I read the results over and over. My mind became a blur of possibilities. One of the NSSL scientists confirmed that the results were real, but it was obvious the installation of government Doppler radar was ten years or more into the future. I couldn't wait, and I believed the people of Oklahoma shouldn't have to wait either.

Armed with a few facts, a burning desire, and belief in my ability to accomplish my goal, I picked up the phone and nervously dialed a number in Enterprise, Alabama.

"Hi, this is Gary England with KWTV in Oklahoma City. Is Neil Braswell available?" I asked.

In a slow, southern drawl the lady on the other end replied, "Just a moment, Gary, and I'll check."

Neil Braswell, a bright, electronics engineer, was the owner and president of Enterprise Electronics. His operation was the designer and manufacturer of 90 percent of all radars in the free world. KWTV had purchased two radars from Enterprise Electronics. My experiences with the company, its employees, and Braswell had been very good. If I had any chance of reaching my goal, Braswell and his crew would play a vital part.

"Hello, Gary. What can I do for you?" said Braswell.

"Neil, can you build us a Doppler radar?" He and I both knew that military warning radars had been modified for the purpose of weather research, but we also knew that no one had designed and built a Doppler weather radar for the private sector. What I wanted existed only in my mind.

From Braswell's end, there was a very long pause. "Tell you what, Gary. That thought has never crossed my mind. Let me talk with my people, and I'll get back with you."

As I awaited word from Enterprise Electronics, the spring of 1978 moved slowly. Tornado activity in Oklahoma dropped to lower than normal, which allowed me more time to study the NSSL research results. It became obvious to me that Doppler radar would play a big role in future severe weather warnings. The probability of tornado detection with Doppler was estimated to be 30 percent better than present conventional means. Experiments suggested a dramatic reduction in false alarms. The critical success index was three times better than what was then available in the real world.

I tried to temper my growing excitement. I knew there could be a difference between experimental results and the results that would be obtained in practical application. Nonetheless, even if there were big differences, Doppler would represent a significant advance. I could see it and feel it as I poured over every piece of research data I could lay my hands on. Each day, between storms and studying research, I waited for Neil Braswell to call.

As I agonized in my very private world about obtaining Doppler radar, the world of television went on. Another news director (my sixth) was appointed. He was sharp, knowledgeable, and strong-willed. He had experience and knew what television news was all about. Of course, my opinion of him was enhanced by his high regard for meteorologists. He understood Doppler radar and its potential impact on the viewing public.

Finally, in the fall of 1978, Braswell called. "Gary," he said in his Alabama accent, "we can build you a Doppler radar."

My pulse quickened. "What will it cost, Neil?"

"I don't know, Gary."

"How long will it take to design and build it?"

"About two years."

"Neil, about how much do you think it will cost?" I half-asked, half-pleaded.

At that point, there was a long, long pause. "Well, it will probably be near two hundred and fifty thousand dollars."

"Two-hundred and fifty thousand," I echoed, trying to conceal my fears about how I would ever convince John Griffin to come up with that much money for something that didn't even exist yet.

"Tell you what, Gary, I'll firm up the cost and mail you a quotation." Then he added, "I believe you have your work cut out for you. Good luck in selling Mr. Griffin on this because we're going to need most of the funds up front for research and development costs."

Within a few weeks the letter arrived. Two hundred and fifty thousand dollars, half up front, and a delivery date in the early spring of 1980. My work definitely was cut out for me.

Like a squirrel caught in an early winter, I frantically gathered what was most important, as quickly as possible. I put together a presentation that focused on Doppler radar's potential for allowing a twenty-minute warning lead time. Saving lives was my primary theme. My secondary and supporting theme was on the promotional benefit of owning and operating the world's first Doppler radar dedicated to public warnings. If successful, KWTV would make history.

I had to acknowledge the possibility that the Doppler might not be successful. The NSSL Doppler was backed by millions of dollars for research and development. We and Enterprise Electronics were going to produce one for $250,000. It was an effort that could fail. In that case, my future would most likely not be television.

My mind raced ahead to the future. I might end up back on a farm, breaking ice in the dead of winter so the cattle could drink. I saw visions of myself shoveling truckloads of musty grain in the sizzling summer heat.

I also saw myself mixing barrels of slop to feed squealing, always hungry pigs. I had already done all of those things as a child and teenager, and I had no desire to return to them.

Many of the negative experiences of my life popped up in my mind like images in a photo album. The picture of Christmas with no money to buy my wife a present sprang up from the depths of my memories. A never-forgotten Thanksgiving without any food and less than five dollars in our pockets became real again. The image of driving my wife to work when no one would hire me was real and provoked a powerful, almost palpable fear.

But my desire to operate the world's first commercial Doppler radar was all powerful, all consuming. It overcame my fear of what might happen if the project failed. I knew if KWTV accepted my argument for Doppler radar, I would be operating on the edge until it was proven a success. The great abyss might be out there, but I had to take the chance.

In a series of meetings with Griffin and DeLier, I made my case. Griffin was fascinated by what Doppler radar was and what it could accomplish. He asked very perceptive and intelligent questions about the operation of such a system. He already knew the value to the viewing public of significant lead time on tornadoes and severe thunderstorms. I kept getting the feeling, however, that Griffin had some doubts about the ability of Doppler radar to do what research and experiments said it could do. He was, though, a man of great vision and was willing to take risks. He also kept things in perspective and was fair in his dealings with and treatment of employees. Any business proposal I might make would, without a doubt, receive fair and serious consideration.

About an hour into the last meeting on the purchase of the Doppler, the room became totally quiet. I knew the time had come. I could feel the rapid thumping of my heart. Griffin looked up from his notes and fixed his eyes on me. I felt very uncomfortable.

"Gary," he began, "since you came here in 1972, we have purchased a significant amount of weather equipment. We have spent large sums of money in your area." I wanted to swallow, but my parched mouth and throat refused to cooperate. "You have made," he continued, "many recommendations, and I must say, not once have you mislead me on what this television station needed. We're going to buy this Doppler radar. Jack and I will proceed from here. This meeting is over."

Griffin had just given me one of the highest compliments a person could receive. I sat alone in the room and pondered his words. Acquiring the Doppler was secondary at that moment. He had just told me that he believed in me and trusted my judgment. To this day, his words serve to inspire and guide me through tough times and difficult decisions. He was a true leader.

As I charged down the hall to call my wife with the fantastic news, I once again had the thought: What if they build the radar and it doesn't work? At that moment, I encountered DeLier and received, without asking, my answer. "That damn radar better work, England," he said as he passed by. "It's a long way from the top of our sixteen-hundred-foot television tower to the bottom," he added.

DeLier was, of course, kidding. In television the real threat was that quick trip from the top of the ratings to the bottom, from job to no job. I had seen it happen many times. One moment an anchor would have the world by the tail; the next moment, he or she would be looking for a new job.

An anchor must be successful on the air. In most cases, that requires an acceptable voice, appearance, and the ability to communicate well with the audience. It requires projecting a warm and likable image to not just a few viewers, but to the masses. In television, if people don't watch your show, your job will be given to someone else.

It's not much different from owning a restaurant or being a retail salesperson. If no one eats at your restaurant, you will soon have no restaurant. If you don't make adequate sales, you will soon have no job. In television, if an anchor does not do well enough to maintain or increase the number of persons watching, he or she will be replaced by someone who does a better job. It's a fact of life in television.

There are other threats to an anchor's well-being, but they do not represent a real danger. On stormy days and nights, when the weather seems to cause perpetual interruptions of regular programming, negative calls from a few viewers are almost inevitable. For the most part, the calls come from viewers who are not located in the area affected by a severe thunderstorm or tornado. Frequently viewers wonder why severe weather interruptions cannot occur during commercials rather than the scheduled programs. Such calls can be exasperating because the caller may not appreciate the importance of commercials. Television income is derived from advertising. It is, therefore, impossible from an economic standpoint to eliminate commercials. However, when there is a threat to life or property, commercials are also interrupted by the warnings.

In addition to adequate commercial revenue, a television station also depends on the success of its employees. If the station is successful, jobs are maintained, and the station is an asset to the community and state. If, however, the station is unsuccessful, it will be downsized and its value will decrease.

One thing certain to affect the employees is the appearance of a new news director. Anxiety may be greatest when the position is filled by a person from within the company, especially when that person is one that a few employees may have treated rudely.

During the 1970s the initial staff meeting with a neophyte news director was predictable. The general theme usually was something like this: "I'm here. I'm tough. I will not put up with slackers. The story writing is not professional. Some of the video is downright embarrassing. The teases are worthless. If you mess with me, you'll find yourself looking for a job."

In fact, television anchors are frequently looking for other employment. Sometimes it's their own choice, other times they're forced out, and occasionally they self-destruct. Twice I've seen news anchors tangle with top management in loud, face-to-face confrontations. Both exchanges resulted in rather negative results for the anchors. Quiet observation of such episodes quickly taught me to pursue my objectives through other means.

On occasion an anchor will just disappear from the small screen, never to be seen on television again. Understandably, this upsets some viewers. Their favorite television person is gone, with little or no explanation.

The sportscasters I've known have been fine individuals with great talent. They seem to be truly happy only when watching sports, participating in sports, or doing their on-air shows. Quick-witted and outspoken, sportscasters are a fascinating element of television.

One sportscaster, unhappy with his boss, decided to quit. When I heard of his decision, I went to say goodbye.

"John," I said, "I'm really disappointed that you're leaving."

"Gary," he responded, "if I stayed another minute, I would commit felonious assault." That reply cut right to the chase—a trademark of sportscasters.

In the weather department, new employees are hired to do designated on-air shows, assist the chief meteorologist, and carry out other specific responsibilities. On occasion, though, a new employee is after the chief meteorologist position.

One weather department employee, after just over two years on the job, went to the news director and demanded that he be given my job. "Not in your lifetime," the news director responded. The employee walked out in search of employment elsewhere.

But there was a problem. He still had several months to go on his contract with KWTV. About a week after his departure, he called me. Very apologetic, he gently informed me that God had spoken to him. He had been directed to return to KWTV and finish out the contract. His conversation with God may have taken place, but I knew for sure that our general manager had talked with his new boss. The message was to send him back or face a contract-tampering suit. The employee came back.

Behind the camera, television weather is a real-world, no-nonsense job. KWTV requires exceptionally high standards of work performance, dedication, and personal behavior.

For less than a year, we had on the air a very personable individual. His on-air presentation was very good, but behind the scenes, he was somewhat less than cooperative.

One spring day, severe weather exploded across Oklahoma. In between warnings, I paged him. He did not respond. I called his home. No answer. Finally after about two hours, he called.

"You want something?" he asked in an irritated tone.

"We have severe thunderstorms and you need to come to work," I replied.

"Well, I'm mowing the lawn, and it's not raining at my house," he responded. He then hung up the phone and did not come to work.

Such actions, I knew, could lead to early departure. With prospects of operating the world's first commercial Doppler radar, I made sure my behavior stayed within expected norms. But before the Doppler actually came into being at KWTV, a considerable number of wild events would be encountered.

Chapter 11

The First Doppler

The spring of 1979 was tumultuous. Just after 3:00 P.M. on April 10, a tornado touched down near Foard City, in north Texas. It moved swiftly toward the northeast at 60 MPH. Nine minutes later another savage twister formed and ripped through Vernon, Texas, with winds of 250 MPH.

This was the beginning of a terrible outbreak of long-tracked tornadoes across northern Texas and southern Oklahoma that did not end until after 8:00 P.M. The five hours of horror became known as the Red River Valley tornado outbreak.

Fifty-nine people lost their lives and nearly two thousand suffered injuries. All but four of the individuals killed were in Texas. The small town of Vernon suffered greatly, with ten persons dead and seventy injured; but the worst suffering was reserved for Wichita Falls, Texas. There, a twister over a mile wide and with winds in excess of 250 MPH killed forty-five and injured more than seventeen hundred. Nearly six thousand homes were destroyed, leaving roughly twenty thousand people homeless. Damages amounted to $400,000,000. All of this took about fifteen minutes.

Dave Tomez, a future KWTV employee, was then employed at Sikes Shopping Center in Wichita Falls. Out of the corner of his eye, he saw people running through the mall. Above the commotion caused by hundreds of people caught in the grip of panic, he heard, "Tornado!".

Tomez, a curious sort of fellow, decided to go with a security guard out of the mall. They wanted to see exactly what was happening. Much to their surprise, they saw the tornado moving directly toward them, devouring chunks of the shopping center. Tomez and his companion were so close to the twister, it looked like a huge, boiling, horizontal tube.

At that point, no instructions were needed. With his heart pounding like a drum, Tomez ran back into the mall. With the demon right on his heels, he dove headlong into a recessed seating area already covered with people. He believes to this day that what saved him from serious injury was another person who landed on top of him, shielding him from the barrage of debris. He did not escape unharmed, though. Hundreds of slivers of glass were stuck in his head and torso.

The warning lead time on the Wichita Falls tornado was considerable. The experimental Doppler radar at the NSSL suggested the possibility of a tornado sixty minutes before the storm hit. On-the-scene storm spotters followed the developing storm for over an hour as it approached the hapless city. Everyone involved in emergency management knew the tornado was headed toward Wichita Falls. The warning sirens sounded time after time. Radio and television repeatedly broadcast the approaching threat. Still, the twister took a dreadful toll in lives and property.

More accurate, earlier advance warnings are never going to preserve property, but in theory, they should allow most everyone to move from harm's way. In Wichita Falls, with the large number of fatalities, it appeared something was terribly wrong with the theory. In reality, though, some people will not hear the warning that a tornado is on the way. Furthermore, not all of those who hear the warning will know what to do. Also, when faced with an oncoming twister that may represent the end of life, human reaction is going to be unpredictable in many cases.

At Wichita Falls, of the forty-five persons killed, twenty-five died in vehicles—nearly 56 percent. Nine deaths occurred outdoors in the blinding torrent of wind and debris. Eight others died in homes or public buildings. Three residents suffered fatal heart attacks.

Of the twenty-five fatalities that occurred in vehicles, sixteen were people who saw the tornado approach and then decided to use their vehicle in an attempt to avoid the tornado. Damage research revealed that eleven of those sixteen left what turned out to be the secure shelter of their homes and businesses. They probably would have survived had they stayed where they were.

Using Wichita Falls as an example of what happens when a large tornado strikes a densely populated area, the results suggest that many will die, regardless of advance warning. In addition, the Wichita Falls disaster indicates quite clearly that vehicles offer no protection in a tornado and may well, in many cases, produce the majority of deaths. The monster Wichita Falls tornado raised everyone's awareness to new heights.

Finally the spring of 1979 slipped into summer, but with it came thirteen tornadoes in August, the most ever recorded in Oklahoma for that month. It felt like the storm season with no end. Long hours at the radar combined with communications traffic from the field and numerous confrontations with new, inexperienced producers tested my patience. Eventually, though, the power-generating heat of summer gave way to fall, and normal weather returned.

With the fall season, came a significant change at KWTV, a development that would influence Oklahoma television for many years to come. A new person was hired in top management. At that moment, I had no idea that my job and future were at risk.

Bob Holding, a very successful KWTV salesman, invited the new man to dinner. After dining, Holding turned on the 10:00 P.M. news. After about ten minutes, the weather came on. There I was, with long unruly hair, a weird suit, and flashy boots.

The new man, pointing toward the screen, turned to Holding and said, "That's the first one we have to get rid of."

"Well, Duane," Holding said, "I don't know quite how to tell you this, but that little guy there doing the weather is the most popular talent in the market."

Duane Harm, the new station manager and an exceptionally bright and knowledgeable businessman, was taken aback by what he had just heard. Reviewing my tapes from that era, I can certainly understand his analysis of the situation. He was, however, analyzing and planning even more than any of us expected.

Within two months, a distressing message was delivered to me by the news director. Harm was reviewing expenditures and was planning to cancel the purchase of the Doppler radar. This news shook me. I knew that Doppler radar was the future. After requesting, then pleading, and finally begging, I was allowed a one-time opportunity to convince him that the order for the radar should stand. I was given twenty-four hours to get my plan together.

In a frenzy, I gathered the research statistics on warning lead time. Then I scooped up the few color reprints of Doppler presentations that were in my possession. I had one day to save my dream. I located and talked with an employee who was good with a 35mm camera. He agreed to take slide shots of the few pieces of data that I would use to support my position. By late afternoon, I had persuaded someone to develop the film and have the slides ready the next morning.

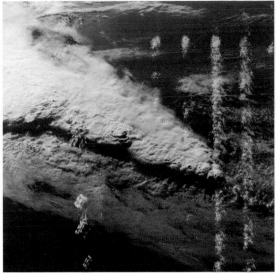

Above left: Space-shuttle view of thunderstorms over the Gulf of Mexico (Photo: NASA, David Pitts)

Above right: Line of thunderstorms as seen from above (Photo: NASA, David Pitts)

Left: Rotating thunderstorm that produced baseball-sized hail near Ada, Oklahoma, April 29, 1978 (Photo: Don James)

Significant tornado near Wakita, Oklahoma, April 1991 (Photo: Linda Elaine Krejsek)

Cumulus congestus clouds indicate atmosphere is becoming unstable. (Photo: Martha Carrol)

Mature thunderstorm with a back-sheared anvil and significant flanking line (Photo: Sue Sadler)

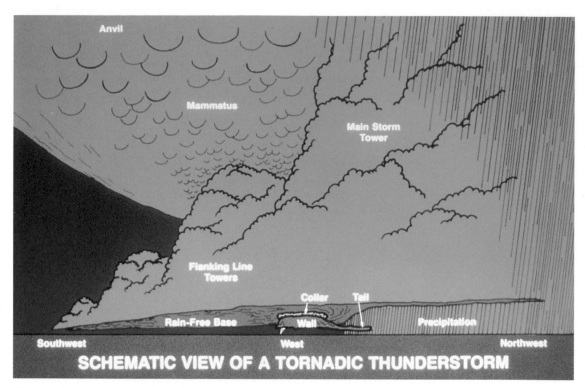

Idealized side-view schematic of a tornadic thunderstorm (Photo: NOAA)

Side view of a severe thunderstorm with a wall cloud (Photo: Sue Callahan)

Rotating wall cloud near Ringwood, Oklahoma, May 30, 1991 (Photo: Dwight "Ike" Gauley)

Rain-wrapped wall cloud over Oklahoma City, March 28, 1988 (Photo: Rick Foster)

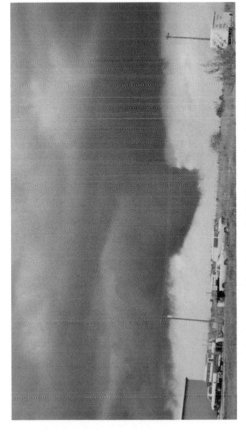

Rotating wall cloud approaching Edmond, Oklahoma, May 8, 1986 (Photo: Doug Fletcher)

Rotating wall cloud over Enid, Oklahoma, June 14, 1986 (Photo: Tom Meyer)

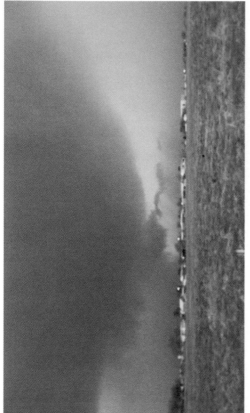

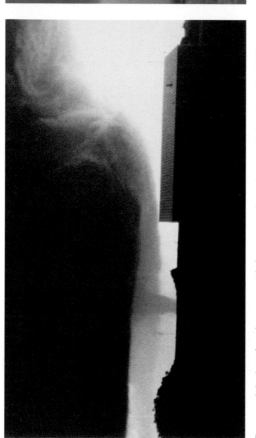

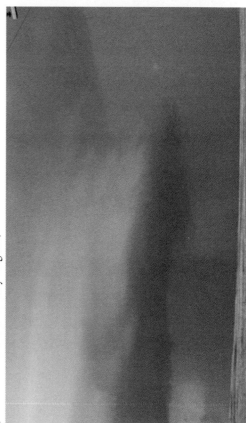

Above and below: High-precipitation rotating thunderstorm with wall cloud over southwest Oklahoma City, March 28, 1988 (Photo: Chris Floyd)

Low-precipitation thunderstorm with large tornado and no wall cloud, north of Clinton, Oklahoma, May 22, 1981 (Photo: Susan Muno)

High-precipitation thunderstorm with no wall cloud and with tornado obscured from view (Photo: From the files of Gary England)

Above and below: Lightning over Oklahoma City, July 1987 (Photos: Tony Bennett)

Tornado revealed by lightning near Miami, Florida, June 13, 1985 (Photo: James Leonard)

Lightning usually strikes at the highest point, but not always. July 1987. (Photo: Michael W. Smith)

Above and below: Golf-ball- and larger-sized hail, Bray, Oklahoma, March 28, 1988 (Photos: George R. Noe)

The atmosphere produces many rotating tubes. This one is horizontal, approximately five miles wide, and more than one hundred miles long. Northwest of Hawaii, October 5, 1985. (Photo: NASA, David Pitts)

Long, ropelike tornado near Hennessey, Oklahoma, April 1990 (Photo: Bob and Marie Stewart)

Weak funnel originating from a fair-weather cumulus cloud (Photo: Tye Smith)

Two small funnel clouds connected to a flat-based thunderstorm, near Woodward, Oklahoma, April 9, 1992 (Photo: Jim McCormick)

Small tornado with dust wrapping around the circulation, near Woodward, Oklahoma, April 9, 1992 (Photo: Jim McCormick)

Above and below: Two of the twelve tornadoes that formed near Woodward, Oklahoma, April 9, 1992 (Photos: *above,* Don Downey; *below,* Val Castor)

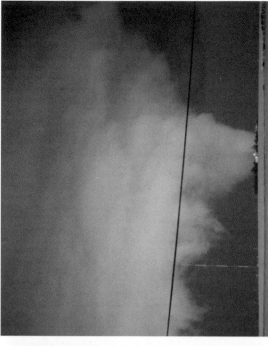

Above: Anticyclonic tornado near Alva, Oklahoma, June 6, 1975 (Photo: James Leonard)

Above right: Large multivortex tornado near Pampha, Texas, May 19, 1982 (Photo: James Leonard)

Below right: Pampha, Texas, tornado, May 19, 1982 (Photo: James Leonard)

Cone-shaped tornado northeast of Enid, Oklahoma (Photo: Nick Lohse)

Wedge-shaped tornado near Weatherford, Oklahoma, April 22, 1985 (Photo: Kelly Fry)

Small tornado originating from a benign-looking cloud, northeast of Clinton, Oklahoma, May 22, 1985 (Photo: Jess Karber)

Tornado with winds of 160 to 200 mph in Edmond, Oklahoma, May 8, 1986 (Photo: Dan Roberts)

Above and below: Violent tornado passing within one mile of camera, near Red Rock, Oklahoma, April 25, 1991 (Photos: Richard Oltmann)

Damaging tornado in Spearman, Texas, May 31, 1990 (Photo: Jeanie Dearey)

Tornado near Hinton, Oklahoma, May 6, 1990 (Photo: Jerry Boling)

Tornado in Ada, Oklahoma, March 15, 1982 (Photo: *Ada Evening News,* Steve Boggs)

KWTV Doppler radar velocity presentation of mesocyclone that produced tornado that hit Ada, Oklahoma, March 15, 1982 (Photo: KWTV)

Above and below: Violent tornado moving through Heston, Kansas, March 13, 1990 (Photos: Emily Laughlin)

Destroyed home, Heston, Kansas, March 13, 1990 (Photo: Emily Laughlin)

Example of what 300-mph winds will do to a vehicle, Heston, Kansas, March 13, 1990 (Photo: Emily Laughlin)

Above and below: One of two tornadoes that just missed Clinton, Oklahoma, on May 22, 1981. A third tornado caused significant damage. (Photos: Darrel Moser)

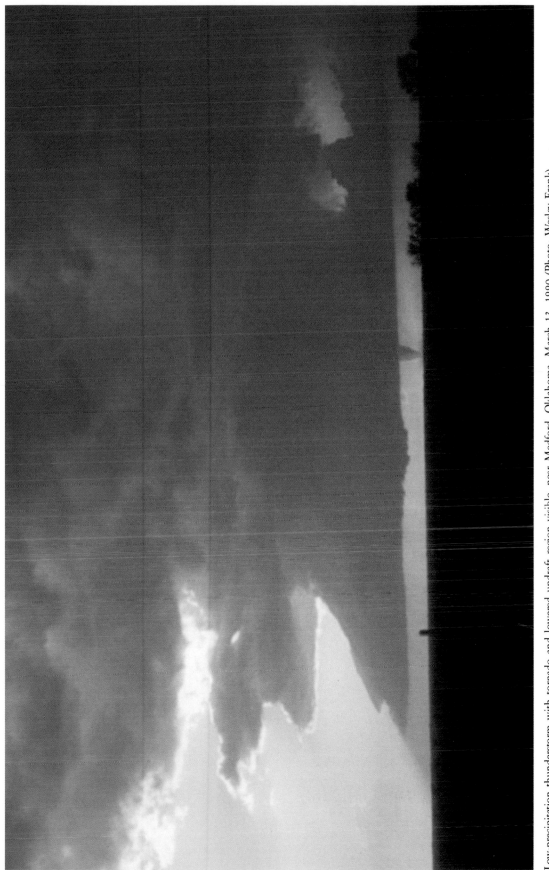

Low-precipitation thunderstorm with tornado and lowered updraft region visible, near Medford, Oklahoma, March 13, 1990 (Photo: Wesley Frank)

This sequence of four photos shows a violent tornado with winds of approximately 300 mph crossing Interstate 35 north

of Perry, Oklahoma, April 26, 1991 (Photos: From the files of Gary England)

Rotating low-precipitation thunderstorm with large tornado, near Laverne, Oklahoma, May 15, 1991 (Photo: James Leonard)

Union City, Oklahoma, tornado of May 24, 1973 (Photo: Joe Demmer)

Sleep was difficult that night. I knew the next morning would bring a tough confrontation with a person who seriously doubted the usefulness of spending a large sum of money on something he had never heard of, and considered of doubtful value to television. As the first rays of sunlight split the early morning darkness, I was still awake. I hoped for the best but expected the worst.

In the news director's office, I placed the projector on a slightly tilted chair seat. I tried to level it. It remained crooked. A rough, pale green wall substituted for a screen. With the news director hovering close by, Duane Harm came into the room followed by a few designated observers. I knew he didn't care for me much, and his stern expression didn't help matters at all.

The room was very quiet. On the floor, I knelt beside the projector. Up popped the first slide. I started the sales job of my life. By the time the tenth and final slide came up, my knees hurt and my voice shook. "Gentlemen," I said, "the television station that first obtains and uses Doppler radar will make history, save lives, and develop a number-one reputation that will be unbeatable." With that final statement, I struggled to my feet. My left knee popped loudly in the silent room. There were no questions. Just, "Thank you, Gary, We'll let you know."

I stared blankly at the group as they left the room. I had never been in a meeting where there was no reaction at all. I had no idea what the decision would be. But I knew John Griffin was the man at the very top, and he wanted the Doppler radar.

By the first of the year, I was notified by an assistant to an assistant to the news director that the purchase order for the Doppler radar was firm. The due date, however, for the radar came and went. With no Doppler and no word from the manufacturer, I was a little more than concerned. I called Hal Quast, then my primary contact at Enterprise Electronics.

"Hal, this is Gary. What's the status on the Doppler?" I asked in a friendly but worried tone.

"Ah, Gary, my boy," he said in his standard response to all of my calls, "how is the world treating you?"

"Doing great, Hal, but what about the radar?"

"We're a little behind schedule."

"Hal, what does little mean?"

Quast, a very likable person with a lanky, string-bean build was one of several engineers working on the project. He was a straight shooter, but I wasn't prepared for the bullet he was about to fire. With a very jolly, damage-control voice, he said, "About a year."

His message hit me like a sonic boom. With the promises and claims I had made to management, I knew that a year delay meant I might not even have my job at KWTV when the Doppler finally arrived.

"Hal, not a year!" I demanded.

"We'll have the system in your hands by next spring," Quast clearly and precisely informed me. "That's the way it is, and I can't change it."

Fortunately, due to a very quiet storm season, little attention was focused on the weather department during 1980. There were only twenty-five tornadoes and seventeen injuries in the entire state. It was definitely a smooth year in weather, but still a very turbulent period within the brick-and-steel confines of the television station. Also, news director number seven had joined the staff.

"There will be no significant changes." When I heard those words early in 1980, I suspected that major alterations would soon occur throughout the company, ripping apart the infrastructure to which we were so accustomed.

New faces appeared as old ones disappeared. It was obvious that the search for quality and new loyalties was underway. Some employees were demoted, others were promoted. No department was immune. From sales to engineering to marketing to news, the fine-toothed comb of change swept the company, leaving no one untouched.

Some departments were renamed, others were reorganized, and still others were restaffed. Everyone felt nervous about the future. A few employees left the business forever. Others fought valiantly for their jobs. Some succeeded while others did not. The waves of change lessened with time, but they would continue to rock our world for nearly ten years.

For the on-air talent who survived, some of the changes were good. Professional makeup persons were brought in to improve our pasty, embalmed appearances. The station manager's wife became our fashion and design consultant. That offended a number of people, but she was good at her job. That became evident as we gradually acquired a sharper professional look on the air and off. Male anchors were ordered to wear four white shirts and one blue shirt during any five-day period. Dark dress suits became the absolute rule. All ties were to be tied with a Windsor knot. Our fashion consultant even got me out of my recently acquired cowboy boots—a major accomplishment in itself.

Even positive changes can be difficult. It became obvious that transforming an operation with a twenty-seven-year history of doing things a certain way was not going to be easy. Success, though, would be achieved.

Through this major upheaval, some managed to keep their sense of humor, while making it known their disdain for news and weather anchors. On the newsroom bulletin board, for example, someone placed a "doctored" photograph of two very ugly old men with thick glasses, missing teeth, and messy hair. A creative caption had been attached to the picture:

Jerry Adams and Gary England are back together again. It was ten years ago today when KWTV management let them go. A survey conducted at that time revealed that their entire audience consisted of ward five at the Central State Mental Hospital. But, why are Adams and England smiling? After leaving KWTV, Adams made a fortune as a denturist (you can see some of his outstanding work in his own mouth), and England finally got a $15 royalty check from his book, "You Too Can Make a Living Forecasting the Weather Inaccurately."

January 1981 finally arrived, and along with it came news director number eight. The due date for the radar was at hand, but so was the possibility that I would follow the trail of those who had left KWTV. I knew the station manager still didn't care much for me, but I didn't know the extent of his feelings until the newest news director summoned me. I settled uneasily into a hard, straight-backed chair.

"Everything in this business eventually changes," the news director said in his most fatherly voice. "You," he continued, "have outlasted most television series, and that too will change." He then paused and blessed me with a long, hard stare.

At this point, I thought I was going to be fired. I sat there, trying to look tough, but unable to utter a single word.

"That's all I have for now. You can go back to work," he stated matter-of-factly. Relieved, angry, and frightened, I returned to my office.

My world became even more unstable when, a short time later, a friend informed me that a member of the KWTV management staff was on the road, watching weather shows on television stations in other towns. He was looking for a new chief meteorologist.

An old, and true, saying came to my mind: "Action cures fear." I tracked down the manager and called him at his "secret" location, a hotel room in New Mexico. I had expended a great deal of effort in finding him. The surprise in his voice when he heard mine was worth it all. He denied the objective of his mission, but at least he knew that I knew what he was doing. I don't believe my action saved me. I believe that John Griffin, again, came to the rescue.

Much to my surprise, I was still employed when the first Doppler radar in the world dedicated to public warnings arrived at the television station. It was early May 1981. The tornado season in Oklahoma had already resulted in the death of five persons in the Tulsa broadcast area and one in the Oklahoma City region. Before the season ended, seventy-six tornadoes would sweep the Oklahoma plains.

The wooden crates that shielded the precious Doppler radar components looked beautiful. What was in the crates, however, looked like useless clumps of electronics. After what seemed to me like hours of circling the boxes, the engineers finally moved in and began extracting the various components. The chief engineer eventually told me he thought everything would go more quickly and smoothly if I would please get out of the way and let them do their work. He was correct.

With Doppler installation underway, the selling of the concept to the public began. A torrent of television, radio, and print advertising poured forth. The phrases "In twenty minutes Oklahoma City could be destroyed" and "Now you could have an extra twenty minutes to live" were the marketing themes. I was aghast. We seemed to be promising a twenty-minute warning on all tornadoes, an impossibility then and now. The marketing approach was quickly toned down after I convinced those involved that, in truth, we *should* be able to provide a twenty-minute warning on *some* tornadoes. A softer approach, "Doppler StormScan sees the wind," was adopted, but it was hard to shake from the advertising tree the "We'll save your life and no one else can" message.

The advertising campaign was very effective. Along with my on-air enthusiasm about the marvels of Doppler radar, it had a significant impact. The people of Oklahoma became very excited about the prospect of better warnings. Some, however, at the NWS and the NSSL were not so pleased. It was only because of the success of the NSSL's publicly funded research that we were able to acquire—for commercial purposes—the world's first operational Doppler radar dedicated to public warnings.

Some of the national media picked up on the significance of the private weather movement in Oklahoma. The May 29, 1981, issue of the *Wall Street Journal* stated, "While the government cranks up its ponderous procurement plan, others aren't waiting. Station KWTV in Oklahoma City, where thirty-four tornadoes have struck since 1892, is installing a new Doppler radar for issuing its own warnings to viewers." That was great publicity for KWTV but added to the friction between some government meteorologists and myself. It would be twelve long years, 1993, before the first National Weather Service Doppler radar was officially commissioned and accepted for operational use. During those years the government system was called NEXRAD, an acronym for "*nex*t generation *rad*ar." Once commissioned, the name changed to WSR-88D or, as it is usually referred to, Doppler radar.

Late on the evening of May 14, 1981, the KWTV Doppler was up and running. Immediately, just to our west, we saw a large blue-colored radar return moving toward the station. Something was wrong. Employees looking out the back door could not see a single cloud in the sky. I was sure my worst fears were being realized. They had built it, and it didn't work. Hal Quast, the attending Enterprise Electronics engineer, smiled and said, "That blue blob you see represents a large area of insects." Initially, I thought he was just trying to cover up a serious defect in the product, but he wasn't. Doppler radars do reveal concentrations of insects, birds, and bats. But I wanted to see thunderstorms. My wait would be a short one.

"Priority one, tornado warning, take pointer over radar!"

"Priority one, tornado warning. Are you ready?"

"I am ready!"

"We're in the beeps. Cut it and go!"

It was late afternoon on May 17, 1981. The all-too-familiar call to battle was automatic but still a spine-tingling experience.

A small, short-lived tornado quickly sliced through an area five miles southeast of Norman. Another rapidly intensifying storm was also making its sinister presence known. The second storm was very close to the television station. Ear-splitting thunder was continuous. For the first time, I was happy I had no window.

I looked at our conventional radar. It displayed a highly suspect severe thunderstorm that possibly contained a tornado. A small hook echo was evident. A bounded-weak echo region was also present. In addition, the top of the storm was directly over the weak-echo region.

I looked at our new revolutionary Doppler radar. The presentation looked like scrambled eggs. I had never in my life seen anything that resembled the mess displayed on the "miracle radar." The Doppler view appeared to be the ultimate manifestation of absolute chaos. It was impossible to figure out what was happening.

"Whose idea was this?" I asked a little louder than I realized. From the mumbling mob of news people gathered behind, over, and around me in the tiny radar room someone said, "If we had spent the money on cameras, at least we would have known how to use them."

I cleared the radar room. As I slammed the door shut, it really hit me for the first time: no one had ever used a Doppler radar in a real-world application. I was undoubtedly headed for a deep, dark plunge into training myself.

That first day felt like I was on a rapidly moving roller coaster with a broken wheel and no brakes. With every sweep of the Doppler radar, the multicolor display revealed unfamiliar patterns. Eventually, the problem became obvious. The Doppler radar images presented in the radar conference publications had been ideal cases, visual wind patterns that were easy to recognize. When attempting to sell an idea, no one would reveal the messy details of a project. The sample images displayed only clean-cut, marketable results.

Every few minutes, the news director came in to ask which color meant what on our new life-saving tool, assuming I actually knew the answers. The general manager also slipped in for a brief visit. I think he knew I didn't know what I was looking at, even though I responded in professional terms and phrases designed to camouflage my state of shock and bewilderment.

Despite the confusing rainbow of colors on the Doppler, I did learn a little. I had no choice, as over fifteen tornadoes fractured the landscape during a span of five hours. One of them, eight hundred yards wide and with winds of 250 MPH, was on the ground for thirty-five miles.

Bill Jenkins, then a KWTV camera operator, was at home when he saw my first warning for the tornado south of Oklahoma City. Jumping up from the couch, he ran to the kitchen and said to his wife, Barbara, "Honey, would you like to go get some ice cream?" Surprised and delighted that her husband had suddenly become so unusually gracious, she quickly accepted.

In a matter of minutes they were moving swiftly eastward along Interstate 40. Swerving wildly through heavy traffic, Bill had already calculated that he could intercept the tornado near the town of Tecumseh. By this time, Barbara had figured out that they were not going to get ice cream. She was not pleased.

Just east of Oklahoma City, Bill and Barbara cut south on Highway 177. He could see a large thunderstorm to the southwest. Pressing the accelerator to the floor, the KWTV van screamed south. The closing distance between them and the storm decreased rapidly. Within ten minutes, a heavy curtain of rain swept across the highway. At full speed they plunged into the wall of water. The visibility was near zero.

Strong winds buffeted the vehicle. The brief pauses between crashes of thunder were filled with the pounding roar of hail as it bashed the thin metal shell that served as shelter from the storm. Bill and Barbara were penetrating the core of the storm, an experience that even the most experienced professional storm chasers usually try to avoid.

Through the thinning sheets of rain and large hail, Bill looked to the southwest. "Wall cloud!" he shouted. Sliding to a bumpy and sloshy stop, Bill leaped from the driver's seat into the vicious storm. As the core of the storm passed, lightning, rain, and hail decreased but did not stop. Bill quickly extracted his video camera from its case.

A debris plume exploded from the ground upward. "Here it comes!" Bill shouted to Barbara, who was not at all comfortable with their exposure to the advancing tornado.

"Bill, get in this truck!" his wife commanded. The thunder crashed, and hail and rain plastered Bill. Undaunted by the storm or his wife, he continued to roll video.

"Look at it! Look at it over here! You got it?" Barbara yelled.

"I got it, here it comes!" Bill responded.

"There are two of them, Bill. Here it comes. Oh _____, Bill!"

"What?" Bill gruffly responded.

"Bill, get in this truck!"

"It ain't coming this way," Bill snapped, and then continued, "Call . . . call the station! Tell them I've got one on the ground, on video!"

"Bill, get in this truck!"

Recorded on video, it was a real-life drama between wife, husband, and tornado. Several months after the event, Bill mentioned they never did get any ice cream that day.

That was a successful day, with good warnings and great video. There were no fatalities in our viewing area. Eight injuries, though, were recorded—six of them in the twister that Bill had captured on video.

As soon as the storms were gone, I began to worry about how to utilize the Doppler. In every idle moment my mind was flooded with bizarre Doppler radar displays. By day I worried. By night I dreamed. Finally, I forced myself to accept the premise that I would have enough time to gain the necessary experience. In reality, what was in store was a huge amount of experience crammed into a very short time.

Chapter 12

Major Assault

Detailed analysis of severe weather parameters on the morning of May 22, 1981 suggested the possibility that dry-line thunderstorms would form in western Oklahoma during the afternoon. The KWTV helicopter, Ranger 9, was put on standby. Additional reporters and photographers were called to the station. The production manager at that time, Jerry Dalrymple, brought in his best audio person and best director.

Dalrymple, was a hulking man with gorilla muscles and, when necessary, a personality to match. No one in his department dared challenge his orders or authority. He had combined the corporate frown and scowl into an effective management tool. I had seen him stare down some pretty tough dudes. But he was good at getting the job done—there was no one better. And on a day like May 22, 1981, the best was required.

The radar engineer was informed of the weather situation to make sure he didn't leave town. Once, during an outbreak of severe weather, the designated radar engineer happened to be on a weekend outing. He was 120 miles from Oklahoma City when the radar system crashed. No one at the television station knew where he was. After a two-hour search for his location, the Ranger 9 helicopter retrieved him and delivered him to the waiting arms of the station manager. Since that day, knowing the exact location of the radar engineer has been a prime objective.

After I had mulled over the weather situation as long as possible, I said, "Smitty, I would like for you and the news crew to take up a position on Highway 152, just east of Binger." Carl Smith, the microwave truck engineer, did not care much for tornadoes, and thought even less of deliberately exposing himself to one; but off he went on that fateful Sunday afternoon, trailing close behind the news crew.

By late afternoon, the western Oklahoma sky blossomed with towering, mushroom-shaped clouds. The dry line was erupting. Within an hour tornadoes were touching down near Cordell and Lake Valley. Damage was reported. Ranger 9 was launched westward into the fray. It was pressure-cooker time again. Activity in and out of the radar room was at a furious pace.

The exhilaration of dealing with tornadoes on the move is almost indescribable. I don't like what they do, but I love the challenge of trying to anticipate what is going to happen next and where. It produces a high-voltage work atmosphere. With life in the balance, discussions on what to do are heated and loud. A severe weather event that lasts a few hours is a wild trip that leaves a person emotionally, physically, and mentally drained.

At 5:10 P.M., another tornado spun up. While it was on the ground, another, larger twister plowed across the open fields. Ranger 9 was at the scene, but the pilot didn't know where—he was lost.

"Ranger 9 to five-five-eight. We have a tornado on the ground north of Cordell," the pilot announced in a controlled, solid voice.

I yelled out my standard standby for a priority one. I looked at the radar. There was no storm at Cordell, which is about seventeen miles south of Clinton. I wondered if the azimuth on the radar had some how slipped out of alignment.

"Five-five-eight to Ranger 9. State your location again."

"This is Ranger 9 and we are just north of Cordell. Standby for a live microwave picture."

As the helicopter bounced across the sky, the conversation inside became highly stressed. "Oh, oh, I gotta get some shots!" yelled the photographer.

"Zoom back on the shot!" the reporter shouted, repeating the instructions I had just given him.

Ranger 9 shook violently as it strained against the powerful storm just to the west. The three people in the chopper were having the ride of their lives.

In a few moments we were ready. We were going to broadcast live video of a tornado from a helicopter. It was the first time this had ever been done. I still did not know if the pilot was lost or the radar had malfunctioned. The on-the-air sequence, however, was in motion and we were going live, no matter where the helicopter was located.

Before I could utter another word, I saw myself on the on-air monitor. "We have a tornado on the ground in western Oklahoma. Ranger 9 has this storm in sight. Let's go to the live video from Ranger 9," I directed.

Up popped swaying images from Ranger 9. They indeed had captured a large tornado, live and up close. It was fantastic. Mixed with the rhythmic, whooshing chop of the blades was the voice of the reporter, trying to describe the situation.

In a matter of moments, Ranger 9 was caught in the inflow of the tornado. The pilot applied power to escape the clutches of the storm. Turbulence increased. The photographer did a splendid job to keep the twister in focus and in sight. As they were pulled closer to the circulation, muffled yells of excitement and fearful swearing could be heard.

"Let's get the hell out of here, man! Let's get the hell out of here, man!" The roar of the helicopter engine and wind combined to obscure nearly all sounds except the loudest shouts.

"We gotta go! We gotta go!"

"We're caught in the inflow, man!"

For several minutes we continued to broadcast the live video as I talked about it. There were thunderstorms all over western Oklahoma. Unfortunately, there were no storms at the stated Ranger 9 location. I was, therefore, forced to speak, repeatedly, about a large tornado on the ground "out there" in western Oklahoma.

Some of the storms were coming into Doppler range. It was going to be an afternoon rich with targets, with Doppler radar presentations that would add to my limited knowledge of real-time pattern recognition. At least that's what I was expecting.

During my voice-over of live video of the lost tornado, the phone rang. Remaining on the air, I picked it up. A voice on the other end immediately said, "Say, that tornado you're showing on television right now is in a field just west of my house. And it's big. And by the way, I live near Arapaho, just north of Clinton."

The mystery was solved. The Ranger 9 heliocopter was twenty miles north of where the crew thought they were. By then, the chopper had veered away from the storm and lost sight of the tornado.

For over a minute, communication with Ranger 9 ceased. The pilot's skills were being given the ultimate test. The radar room was quiet. We thought they had crashed.

"Ranger 9 to five-five-eight, we're all right now."

I was relieved they were safe. I was relieved that the radar wasn't broken. I was a little frustrated at having broadcasted the first real-time video of a tornado live from a helicopter without knowing the actual location.

The twister was the first of the two that nearly hit Clinton that day. Later, a third tornado would not miss its target.

Not long after recovering our contact with Ranger 9, I heard the director's voice the intercom. "England, turn the Doppler off."

"What!" I bellowed.

"The chief engineer just called. He said the FCC [Federal Communications Commission] temporary license to operate the Doppler expired last week. It has to be shut down or we could lose our station license to broadcast anything."

"No! It can't be!" I shouted.

"Shut it off, Gary. Station manager's orders."

Angry but too busy with the existing storms to express it, I powered the system down. I had missed a golden learning opportunity.

Another tornado was touching down near the town of Swan Lake. At the same time, a monster thunderstorm cell west of Binger was growing rapidly. Our ground crews east of Binger were set up and recording. To the west, the video revealed a rotating, low-precipitation thunderstorm with a huge wall cloud. To the southwest, a sizable dust swirl was evident beneath the rotating wall cloud. The crew members—Phil Van Stavern, reporter, Robert Miller, photographer, and Carl Smith, engineer—were about to meet a tornado over a mile wide. As the storm approached, Smith retreated to the microwave truck, while Van Stavern and Miller innocently faced the oncoming powerhouse.

The sky to the west grew more ugly. Extreme turbulence and rotation covered the sky. A farmer walked up and offered his comments. "There are two more trying to form up yonder there. This one is trying to drop a funnel!"

The wind made a frightful, roaring sound. A few drops of rain flashed through the air, stinging exposed skin.

"Wind is changing," Van Stavern observed in his normal, monotone voice. "We're going to get wet. Get the ponchos," he instructed Miller, a newcomer to Oklahoma.

The wind grew louder. "Get what?" responded Miller.

"The plastic ponchos," came the barely discernible reply.

A new lowering in the clouds was developing to the southwest. The wall cloud was rotating faster. The winds had reached gale force.

On conventional radar, a classic hook echo revealed itself. It was huge and perfectly shaped. Tornado occurrence was likely. A tornado warning was quickly issued for Binger, northeast through Canadian County.

I punched the intercom. "Jerry, you should come look at this! It may be a once-in-a-lifetime chance." Dalrymple, the production manager who had taken over the director's duties, was in the radar room in less than ten seconds.

Looking over my shoulder, Dalrymple saw the unbelievable radar presentation. He reared back, fell over a chair, and shouted, "Tornado! Tornado!" By this time in my career, I was a little more used to such outbursts, but he scared the daylights out of me.

Information from the NSSL chase team, the KWTV chase team, and the radar all confirmed that indeed a tornado was imminent.

In the darkened radar room, I watched the evolution of a feature I had never seen before and have never seen again. On the west side of the huge hook echo, a smaller hook appeared. This means that, on conventional radar, I was looking at the actual tornado. It's unheard of to see actual tornado circulation except on Doppler radar. The smaller hook echo was migrating along the southern edge of the larger hook echo. The larger hook echo, being produced by the mesocyclone, was in turn producing the smaller tornado vortex signature. I began to broadcast continuous warnings on this extremely dangerous storm.

"Oh, man, this wind is strong," Van Stavern observed. You could hear the strain in his voice.

"Dang, this poncho is choking me!" Miller shouted above the increasing blast of the wind and the violently flapping plastic ponchos.

Just to the west, the wall cloud was rotating rapidly. A large, cone-shaped funnel appeared as if magically at the center of rotation. Debris burst skyward. A car with a broken muffler sped blindly westward on Highway 152, directly toward the tornado. It was then 6:50 P.M.

The noise became deafening. Van Stavern yelled, "You see what I see?"

"Yeah!" came Miller's reply.

"Stay with it! Stay with it!"

The wind tore the poncho from the locking collar around Miller's neck. Smith was still in the microwave truck. It shook continuously, tilting toward the north at an unnatural angle. Smith was positive that this was the big one he might not walk away from.

The tornado became a multiple-vortex storm with winds at the center in excess of 250 MPH. To the crew, it appeared to be close by, as if it were moving through the town of Binger. It was really a mile away, just west of Binger and moving toward the northeast.

The tornado, with its multiple vortices now masked in a mile-wide column of dirt, churned northeastward. It was passing to the northwest of the KWTV team. Finally, the powerful winds decreased, and conversation between the KWTV storm chasers became possible again.

"I've got a lot of dirt in my mouth, Miller," Van Stavern said.

"Kind of breathtaking to see that thing almost come and get you," Miller commented.

"I should have done a stand-up broadcast during the tornado, but you wouldn't have heard anything but my knees knocking," Van Stavern replied. They were finished for the day and very pleased about that it.

The severe tornado continued to move mostly over open land. Seven homes, though, were in its path. They were instantly engulfed and then spewed out in tiny pieces to be gradually reclaimed by nature. Cars, trucks, wheat combines, and telephone poles had been hurled about as if they were toys. Dead cattle had been thrown into trees that were stripped of leaves and bark. Over a mile wide, the tornado was on the ground for twenty-five miles. It was in the class of tornado that kills anyone caught above ground, but fortunately there were no injuries or fatalities. Everyone had taken shelter below ground.

The Nokes family's home was directly in its path. After the tornado passed, all their possessions were gone, but they were alive.

"We were listening to Gary, and we started watching the clouds and saw the tornado coming," Mrs. Nokes said.

Mr. Nokes added, "We saw it come over the hill there, come down on that house. Then we saw it hit that house and it got to about here. . . . We ran to the cellar."

When asked if he had been through this type of thing before, Mr. Nokes answered, "No! We've lost a few cattle, but we've never had anything like this."

Activity in the KWTV radar room continued to be intense and serious. Ten minutes into the rampage near Binger, a second tornado made solid contact with the ground, just west of Clinton. The large funnel, visible for miles, quickly dissipated without causing any damage in the open fields.

At 7:26 P.M., the Binger tornado lifted and headed toward the greater Oklahoma City area, home to nearly one million people. The mesocyclone was reorganizing, as evidenced by the hook echo, which again became extremely well defined. Within six minutes it put down another tornado close to the southern sections of El Reno. It was on the ground for three minutes; it then lifted and continued northeastward toward the Oklahoma City metropolitan area.

"Gary, minor damage being reported at El Reno," someone yelled in a panicky voice. I couldn't wait. "Priority one, tornado warning. I am ready." Within ten seconds I sounded the warning for Oklahoma County. In a matter of minutes, dozens of warning sirens screamed to life. In some areas of western Oklahoma County, the police used their car sirens to warn the public of the approaching storm. Thousands of people took shelter, among them hundreds who flooded public shelters.

For over forty-five minutes before the Oklahoma County warning was issued, I had been telling the public that if this storm continued its track, it would threaten the metro area. Amazingly, management at one of the other Oklahoma City television stations refused to allow the weatherperson to interrupt its broadcast of *The Deer Hunter*. Viewers of that station were alerted to the situation by the multitude of civil defense, police, and fire department sirens.

"England, the station manager just called. He said turn on the Doppler and use it!" stated the assistant news director. "A day late and a dollar short!" I curtly replied. Without an FCC license, I was not about to flip that switch. Besides, the storms were so large and well defined that conventional radar was revealing their secrets without the aid of the Doppler.

The threatening tornado had lifted, but the hook echo was still large and well defined. It was moving past El Reno, up over Yukon, and very close to western Oklahoma County. As a precaution, I changed ranges from 50 miles to 150 miles. I detected a storm developing back toward the west.

"Hook echo approaching Clinton!" I shouted. "Priority one, tornado warning, take pointer over radar! I am ready!" No response.

"Do it now!" I screamed. The possible tornado was very close to Clinton and there was no warning in effect.

"Is anyone home up there?" I yelled at the top of my voice.

"Of course I'm here! What do you want?" came the loud and sarcastic response over the intercom. Under the strain, standard procedures were breaking down, but the warning went out. At the same time, the tornado was on the ground only moments away from its first target.

Ron Williams saw the warning. He grabbed his child, and he and his wife tried to leave their house. "But," he said, " by the time we got to the front door, it was coming across the field right there in front of us, and by the time I came back in and got to the closet . . . it was bustin' the walls . . . walls just cavin' in and the roof flyin'." He added, "As we huddled in the closet, the tornado tried to suck the baby out of my arms. Death, that's all I could see was death." Fortunately, Williams and his family survived uninjured in their house, which was nearly flattened.

Back in central Oklahoma, the large cyclonic circulation was now over western Oklahoma City. It was weakening and losing its radar signature. The hook echo disappeared over Mercy Hospital.

It had been a ferocious day, but the results were good. Considerable damage was reported, but only twelve injuries, all in western Oklahoma. Oklahoma City had, once again, dodged a deadly bullet.

Chapter 13

The Revolution
Begins

A glamorous premier party was held during the fall of 1981. It was to introduce the new fall television shows and new KWTV news anchors to the advertisers and advertising agencies. It also showcased our station manager, Duane Harm, who would soon be promoted to general manager.

The site of the party, a rented theater, was easy to find. Glaring searchlights swept the sky with their intense beams. The Ranger 9 helicopter, positioned near the entrance and bathed in bright lights, suggested quality and strength. Guests were greeted at the door by men wearing tuxedos and women in sleek, black dresses. Food and drink were abundant. I had been to premier parties before, but never one like this.

Finally it was time to start the preview. The news director strode confidently on stage. "Good evening, he said. "I would like to welcome all of you to the KTVY fall premier party." Groans and a wave of laughter swept out from the crowd and over the theater like a tidal wave—KTVY was one of our competitors. After that faux pas, I knew that the news director would encounter rough waters ahead. Harm, the man now running KWTV, had a plan and I was sure it included each employee knowing that he or she worked for KWTV.

A new regime was taking over, and changes were ongoing. Old policies were changed, procedures were modified. The television station was in a state of flux. Our new station manager was tough as a boot, but it quickly became obvious that he knew television and understood what must be accomplished in order to win the television rating wars.

As the evening wore on, the time came for the introduction of the new anchors. I had not seen either of them before that night. In fact, I was not even aware that news anchors were being hired. It was the first time in television that I had ever seen a major personnel change kept secret.

Roger Cooper, the new male anchor, very graciously took the stage and made a few comments. He appeared to be a soft-spoken, professional individual. A little nervous, he didn't smile. Then Patti Suarez was introduced. She quickly took center stage, and I could tell that she was going to be great. She had not only warmth and beauty, she also that unteachable attribute called charisma. She was destined to become Oklahoma's darling.

Duane Harm (who had hired Patti) quickly solidified his position and was soon named general manager, replacing Jack DeLier, the man who had hired me. Harm's goal was to win big in the future. That included quick, effective action to correct glaring problems that currently existed. He immediately began building a corporate infrastructure that would support the weight of his strategy.

Old habits were difficult to change. Many employees felt there was an unannounced singular choice, accept change or quit. Some accepted the new marching orders. Others went elsewhere. A major shift in how KWTV should be run was underway, and there was absolutely no indication the movement would slow or stop.

I was pleased when winter and spring rolled around. The high volume of work associated with that time of year would help reduce my thoughts and comments about the turbulent changes that swirled through our lives.

On March 15, 1982, early season thunderstorms sprang to life. I had no idea that it would be a historic day in the world of television weather.

In the darkened radar room, I alternately observed the conventional radar screen and the Doppler presentation. I became a little uneasy. I could see a very intense storm, about fifty-five miles south of Oklahoma City, near Elmore City. The Doppler showed the classic signature of a circulation, but on the conventional system, there was no hook echo. Of course, many tornadoes occur without a hook echo, and a hook echo may be evident without producing a tornado. The storm was moving east-northeast toward Pauls Valley, a beautiful little town nestled in a green river valley and crisscrossed with historical brick streets.

Soon two circulations were visible. In addition to the one northeast of Elmore City, another mesocyclone developed northeast of Pauls Valley. I could feel my pulse quicken. There were no official warnings in effect.

"Elmore City Police Department. May I help you?"

"Yes, sir. This is Gary England at KWTV in Oklahoma City. I have an indication that there's a tornado just to your northeast. Would you take a look and tell me what you see?"

The moments seemed like hours. "Gary, I can see a funnel cloud in the air!"

Throwing down the phone, I hit the intercom. "Priority one, tornado warning, take Doppler radar! I am ready."

In a flash we were on the air. I was standing full-screen with the Doppler radar presentation enlarged behind me. "This is a tornado warning for you folks from just northeast of Elmore City to Pauls Valley and later Ada,"

I said. The mesocyclone was about thirty-five miles southwest of Ada. It would reach there in a little over an hour.

"I have," I continued, "just called the Elmore City police, and they do have a tornado in the air moving toward Pauls Valley. Once again, the circulation is on the southern tip of the storm on Doppler StormScan."

Pointing at the Doppler colors that identified speed and wind direction in the rotation, I continued, "A mesocyclone is indicated. We have confirmed it with the Elmore City police, and there is a tornado moving up near the Pauls Valley area. We are also on a storm near Konawa, and it also indicates a mesocyclone."

The southern circulation passed just south of Pauls Valley, taking dead aim at Ada. The northern storm splashed a tornado down just northeast of Konawa, in Konawa Lake. At that point, I realized that I had just broadcast the first television Doppler radar warnings for tornadoes and shown the actual circulations.

At 5:57 P.M., the Ada tornado viciously sliced a narrow path sixty yards wide and six miles in length through the northern section of the town. The Brook Mobile Home Park took the brunt of the storm, with fifty-one homes destroyed. In the rubble, emergency teams found one fatality and thirty-six injured.

Ada residents had received a one-hour warning from our Doppler radar. Several subsequent warnings were also broadcast due to our Doppler radar and the NWS. Lives had been saved, but what I learned that day continues to be true: no matter how accurate and timely the warnings, people are still going to die in tornadoes.

The following day, the phone rang. "This is Gary. How may I help you?"

"Gary, this is Don Burgess." Burgess was a friend, an employee at the NSSL. He was deeply involved with the Doppler project at Norman.

"Do you have yesterday's Ada mesocyclone on tape?" he asked.

"Sure do, Don," I responded.

"Several of us from the lab would like to come and view it. Would that be all right?"

"Of course," I replied.

A few days later several scientists from the NSSL arrived at the television station. They reviewed the tape several times, exchanging hushed comments. Burgess finally turned to me and said, "Your Doppler presentation is similar to what we saw on our Doppler." They thanked me and left.

Later, I learned that some members of the local meteorological community had developed a raging fire in their gut. I had dared again to issue my own warning in advance of the NWS warning. Adding fuel to the fire was the fact the warning had been the first real-time Doppler warning broadcast to the public.

My attitude then and now is the same. When people are at risk due to severe weather, it is the responsibility of broadcast meteorologists to immediately alert those in danger. The contention that the media must wait for an official warning from the NWS is, in my opinion, not valid. The management at KWTV was aware of the official discontent with my actions, but they stood behind my policy. Life over arbitrary procedures was and is our position.

Those who favor waiting for an official warning argue that warnings from multiple sources tend to confuse the public. This was said then and continues to be echoed today. I have, however, yet to encounter an unbiased person who has told me of any confusion caused by the multiple-point warning system that has been in use in the KWTV broadcast area for over twenty-two years. In reality, the warning system has fostered healthy competition among all parties, including the NWS. The result has been better warnings and a safer public.

On the average, the state of Oklahoma has more tornadoes per square mile than any other place in the world. Yet, in the western two-thirds of Oklahoma, tornado fatalities are rare and injuries are infrequent. The warning system that has evolved here is the finest in the United States.

Oklahoma County, for example, encompasses approximately seven hundred square miles and has been hit by seventy tornadoes from 1950 to 1994. In those forty-four years there have been just two fatalities and 117 people injured. The proof, therefore, is in the results.

The threat, though, of significant loss of life does exist. There is little that can be done for people who are unaware of an approaching tornado, trapped in their vehicles, or who seek shelter in unsafe structures. And, of course, the occurrence of a large, violent twister with winds of 250 or even 300 MPH will wreak terrible havoc. Lives will be lost and property will be destroyed when such a storm passes through any densely populated area.

On March 31, 1982, the *Ada Evening News* ran a story with the bold headline, "Tornado Warning Policies Changed for Community." The article read, "At a civil defense meeting Tuesday, it was decided to monitor KWTV television and the National Weather Service. If either issues a warning close to Ada or if a tornado is spotted near Ada, the sirens will be sounded."

I gave the newspaper story little thought. I was just doing my job. The Ada tornado and the associated publicity, however, did more than just throw a log on the fire. It was more like gasoline being pumped into a raging subterranean blaze. It was not until a few years later, though, that there was an explosive face-to-face exchange between the KWTV general manager and two representatives of the NWS. We learned then that they believed we had asked the Ada City Council to change their warning policy. NWS employees had been angry for years about something that had never happened.

In the *Washington Post*, dated Saturday, April 27, 1985, a writer put it all in perspective in an article titled, "In Oklahoma, He Weathers Well." Staff writer Bob Levey wrote,

> In May 1982, [Gary England] broke into regular programming to warn southeastern Oklahoma viewers that a funnel was taking

shape there. Twenty minutes later, a twister destroyed a trailer park near Ada, and several dozen people who might have been killed had escaped injury.

England was widely praised for his warning. But he won no friends at the National Weather Service because his bulletin had beaten theirs by fifteen minutes—and he said so, on the air.

England made relations worse last year during the evening news show on his birthday. The anchorman asked him what presents he had received. "A letter bomb from the National Weather Service," England joked. All England will say on the record is that he is "not too popular with those guys." Crawford of the NWS says the relationship—a bit like Oklahoma weather—has undergone . . . unsettled times.

Some local observers see the bad blood as jealousy. Gary is announced as Oklahoma's No. 1 meteorologist, and people try to take potshots at you because of that.

The *Washington Post* story was great publicity for me. It even brought a congratulatory note from David Boren, then U.S. senator and now president of the University of Oklahoma. But the article only stoked the professional fires to greater intensity.

Unfortunately, a degree of animosity continues to the present. In the spring of 1994 a government severe storms forecaster posted his thoughts on a local computer bulletin board.

Please spend a spring in Oklahoma City watching KWTV Channel 9. Gary England has been issuing his own severe thunderstorm and tornado warnings, often independent of the [National Weather Service at Norman] office, at other times merely coincidental, for years and years. Some of the Norman folks can elaborate further, but suffice to say Gary England is not the most popular dude in the Norman meteorological enclave.

The Norman meteorological enclave, though, represents only a minute fraction of the broad spectrum of people that affect the world of a KWTV meteorologist.

The life of a television meteorologist may appear to be easy and calm, except during severe weather, but behind the scenes it can be very tumultuous. At least one surprise a day is the rule.

A telephone call from an irate viewer to management is good for at least a twenty-minute crisis meeting. One viewer, for example, was furious that we preempted his favorite show to broadcast a program on severe weather. That action produced not only a twenty-minute meeting, but a fascinating look at the inner workings of a very angry person.

"Why," he yelled at the KWTV telephone operator, "are you showing that _____ Terrible Twister show tonight?" The operator hung up. He called back. "You tell that _____ to leave the programming alone!" She hung up again. He called back. "Don't hang up on me again! I just want to know what the _____ you people think you are doing!" Again the operator terminated the call. When people swear, she is authorized to terminate such calls. He called back. "I'm going to keep calling and bugging you all day today! Gary probably hasn't changed his underwear in days! All he does is sit in that weather room!" She hung up for the final time. Such telephone calls are, unfortunately, within the normal course of daily events.

One evening, just after my 5:00 P.M. weathercast, I answered the telephone. The caller was a normal sounding lady. In the background, I could hear children. "Gary, I would like to apologize," she said.

"Pardon me," I responded.

"Gary, these children are making so much noise. I know they were bothering you," the voice on the other end explained.

"Madam, I don't think I understand."

"But Gary, couldn't you see and hear the children during your show?"

Flabbergasted, I responded, "No madam, I didn't. I can't see or hear you when I'm on the air."

"Oh," she replied and hung up. The world, without a doubt, contains at least a few strange people.

The short-notice cancellation of a personal appearance can convert a normally pleasant person into a somewhat unreasonable individual. Being up all night covering severe weather is a valid reason for cancellation, but still leaves an organizer without a speaker. And sometimes we inadvertently mess up appointments on our own.

Due to internal mistakes at KWTV, I missed a speaking engagement at an Oklahoma City school—not once, but three times. When I finally made it, the general manager of Griffin Television (KWTV's parent company) and the vice-president of operations came along to help smoothe the ruffled feathers of the teachers and administrators. At the end of my tornado safety presentation, one of the teachers made a few humorous comments and presented me with a T-shirt. Printed on the front of it and marked through with an X were the dates I had failed to show up. On the back, in large type was, "Better Late Than Never." It was not one of my favorite moments.

One bright, sunny Wednesday, another internal KWTV mistake caused me to miss a noon speaking engagement. A year later, they invited me back. Unsuspecting, I happily bounced into the meeting room where the luncheon was being held. I was greeted by five members and forty-five empty place settings. Embarrassed and a little upset, I gave my talk. Later, I was informed that the members of the organization wanted to teach me a lesson. They did.

Chapter 14

The Traveling
Weather Show

The spring of 1982 not only saw the first public on-air use of a Doppler radar for tornado warnings but also marked our conversion from plastic and magnetic weather maps to computers and "colorgraphics." The first electronic marvels were limited in their capabilities and not very expensive. Future systems would be faster, slicker, and extremely expensive.

Most television weather shows run from two to four minutes, but modern graphics systems are capable of producing a brilliant display that could fill two hours. Having so much computer power and so little time causes problems. Many television meteorologists create image-packed presentations that leave the viewer dazzled but confused about the actual forecast.

Late spring of 1982 also marked the appointment of another news director, my ninth in nearly ten years. He was a thin, chain-smoking man who acted tough but was actually a little bashful. I set my posture with him quickly. That kept our confrontations to a minimum. He was from the old school—not too sure about radar or the new, fancy computers.

The arrival of colorgraphics was a significant development. It made the weathercasts more attractive and most likely helped increase the number of viewers. Still, face-to-face weather presentations have always been valuable in building a permanent weather viewer base.

For years, I had been speaking at schools, clubs, churches, civic organizations, and town meetings. I would speak anywhere I could find a crowd who would listen. I just told my story—the life and times of Gary England and how I was finally able to realize my lifetime dream. But a telephone call from Stillwater, Oklahoma, modified my game plan a bit.

A civic club requested that I come to Stillwater and present a video show on tornadoes and safety. I had been showing film, slides, and video to small groups and had even given speeches to as many as five hundred people.

"How many seats in the room?" I asked.

"Seven hundred and fifty," replied Jerry Dalrymple.

"Can't do it," I responded, fearing I couldn't draw so large a crowd.

Eventually, Dalrymple convinced me the room would be full, but as the day of the event approached, I was mentally preparing to deal with the expected failure. "Who ever heard of seven hundred and fifty people showing up anywhere to watch a video on tornadoes?" I asked myself over and over.

The production crew arrived early. They wired the video player to several televisions, set up a screen for my slide series, and hooked up the sound system. Wires and cables were everywhere. Dalrymple and I arrived at 7:00 P.M. Show time was set for 7:30. No one was there except for a few civic club members and the KWTV employees. I went into a cold sweat. No one wants to throw a party and no guests show up.

By 7:15, a few people had straggled into the cavernous room. I was about ready to bolt for the door. Thankfully, at a few minutes before 7:30, nine hundred people poured through the doors and overfilled the room.

The presentation turned out to be a smashing success and was dubbed, Gary's Traveling Weather Show.

With 101 tornadoes blasting across Oklahoma in 1982, there was no time to set up another Traveling Weather Show. One twister with winds of 260 to 300 MPH was in southeastern sections of the state. It was on the ground for fifty-three miles and was one and a half miles wide. In the KWTV forecast center, it was an exhaustive year of long days and nights tangled with frayed nerves and hot tempers. We felt separate from the real world and distant from those who knew us best.

Having decided that I needed a break, top management summoned me. Sitting in the elegant, spacious room were several people, all with a "guess what" expression on their faces. I settled uneasily into my chair.

"Gary," the general manager said, "we would like for you, a producer, and a photographer to visit Mount St. Helens." I stared blankly at my superiors. The mountain was still hot from the eruption on May 18, 1980, and might blow again. "Also," he continued, "it is our wish that you visit some earthquake faults in California. Your objective is a series on how earthquakes cause volcanoes."

"Gary, besides giving you some nice time off, this will make you a more well-rounded, experienced anchor," added the news director.

"Gentlemen," I responded, "I do clouds, not volcanoes." It was a useless comment. They knew I was going, and I knew it, too. In television, when a person is given an unwanted assignment, the unwritten rule is: This is the business you have chosen to be in, so do it.

Also on the trip would be Bill Merickel, our photographer, and Bill Jenkins, our producer. Merickel displayed a quiet excitement about the venture. He dressed in all-black clothing and sported a full beard and mustache. Jenkins was a delightful person. He was a guitar player, singer, and humorist. Jenkins, who liked maximum impact, wore combat boots, tattered levis, a lumberjack shirt, a knit stocking hat, and a scrubby beard.

Off we soared, toward the setting sun and Mount St. Helens. Its catastrophic eruption had devastated 156 square miles. It had killed every living thing in the blast area, including sixty-one people. It was last on my list of places to visit.

We soon arrived in Portland, Oregon, rented a car, and sped across the border into Washington State. Near the volcano, Road Closed signs were in abundance. The powerful explosion had destroyed all of the roads that led up the mountainside.

All except one. Jenkins learned that a single lane road had been bulldozed right to the simmering top of Mount St. Helens. That the road was only for official traffic didn't deter him the slightest. Bright red signs announced one thousand dollar fines and jail terms for all offenders. At 60 MPH, Jenkins swerved the car past the signs. Merickel and I exchanged helpless looks.

The road gradually narrowed until it was barely large enough for our vehicle. The view was awesome. No trees, just ash and blackened stumps. In the distance, past the immediate blast zone, there were trees. They were on the ground, all pointing away from the volcano like a giant compass. It was beautiful in an unearthly way.

I finally thought to look down, straight down. "Jenkins, don't take your eyes off the road," I firmly commanded.

"Why?" Jenkins laughingly asked.

"Man, it's at least two thousand feet straight down about one foot to the right," I responded.

Jenkins nodded and stared straight down the road. His normal humor evaporated. Merickel appeared to have slipped into a meditative state. Jenkins slowed the car to a crawl. The scene was so fascinating, I couldn't help but look down. Every time I did, I felt the twinges of a panic attack.

Finally the road widened, leveled, and veered to the left. Jenkins brought the car to a slow stop. To our left and up was the mouth of the volcano, belching a column of steam or smoke—I wasn't sure which and really didn't care. The situation we were in was totally unpredictable. At least with thunderstorms, a person has some idea about what might happen.

In front of us was a broad area of packed ash. From the ash field small plumes of steam escaped, then quickly disappeared in the very cold air. To our right, below several thousand feet of ash and mountain, was the former Sprit Lake. The view was as unreal as it was beautiful.

"Let's get this shot and get out of here," Jenkins said. Merickel and I didn't argue. We quickly set up the equipment, noting in the process that the ground was quite hot.

We swiftly completed our mission, but not before we noticed a small aircraft flying directly toward us. The airplane passed overhead, turned, and came back.

Jenkins, a former air force man, looked up. "Military observation aircraft. We've been had," he commented. "The good guys will be coming up the mountain very soon," he added.

Jenkins was correct. Before we moved fifty yards, flashing red lights and a well-armed, unhappy deputy sheriff blocked our way. Merickel and I remained in the car, wondering aloud if the television station would pay our bail.

Jenkins talked with the deputy for thirty long minutes. Eventually the deputy ceased shouting, waving his arms and pointing, at a huge sign that said, among other things, Live Volcano. When I asked Jenkins how he had kept us out of jail, he just smiled. He was a true silver-tongued devil. Not only were we free to go, but the deputy showed us a much safer way down the mountain. I was ready to go home, but we were just starting.

We drove hundreds of miles, visited earthquake fault zones, and interviewed dozens of people. Also, one of our tires exploded just south of San Francisco. We were on Highway 101, traveling at an excessive speed. Six lanes of traffic stopped and watched as Jenkins fought to bring the crippled machine to a sliding, screeching, and smoky stop. From that point on, the tiny replacement tire kept our speed reasonable and the angle of our vehicle about twenty degrees off of level.

We left San Francisco on a Friday evening. Jenkins and Merickel were still in the same clothes they had started with. I sat between them in a soiled, wrinkled suit. We looked like we didn't belong on that airplane. About ten minutes after takeoff, Jenkins ripped off his seat belt, jumped into the aisle, and began singing "I Left My Heart in San Francisco."

It was nice to get home. I vowed never to leave again.

Winter passed. With the new year, the news director position was vacant again. A KWTV employee came to me and pleaded that I "put in a good word for him" with top management. I liked him, so I did as he asked. It probably didn't do any good, but he got the job. He was my tenth boss and I thought I had a friend in high places. Within two weeks, however, the "normal" confrontations between the news director and the weather department were underway. Some things never change.

Cautiously, in 1983, we set up another Traveling Weather Show in Stillwater. This time Ranger 9 flew in before the show with my co-anchors. Over fifteen hundred people attended. I commented to Jerry Dalrymple, "I don't understand what is going on, but this thing may work."

We quickly set up four more Traveling Weather Shows, soon to be renamed Those Terrible Twisters. Again, over fifteen hundred people attended in El Reno. Still not convinced of our success, we traveled to Enid.

As Dalrymple and I drove through the Phillips University campus at Enid, we saw hundreds of men, women and children walking along the crowded road. "Well, Jerry," I said, "there must be a basketball game here tonight." He slowly turned his head. He gazed at me with a very sad facial expression and silently nodded his head. "This may be the one where no one shows up," I said. He grimaced at the thought.

As we pulled into our reserved parking spot, I could see that all of those people were going into the building that would house the weather show. Dalrymple gave me a sly grin and said, "Got lucky again, didn't we?"

Inside, nearly eighteen hundred wonderful, smiling faces filled the seats and lined the walls. The fire marshal wasn't happy, but I was delighted. My mother was there. My brother was there. It was a very rewarding experience.

Through the years the Terrible Twisters show has evolved into a major production. It includes a four-hundred-square-foot video screen with first-rate projection systems and backups, a powerful sound system with backup, strobe lights, and storm video guaranteed to entertain and thrill all those in attendance. Putting on one of these shows requires the cooperation of the host town as well as its school system and civic clubs.

The local fire marshal is always present to ensure that only the legal number of persons is allowed into the structure. Firefighters and police officers provide crowd control and security for the anchors. It is a unique operation. The program has been imitated, but never successfully.

To date, we have completed over seventy-five shows. This mix of video, humor and show biz has reached an audience numbering over 250,000. When newcomers to Oklahoma ask, "How could this be?" I say, "I really don't know."

On finishing a Terrible Twister show at Woodward, 125 miles from Oklahoma City, we jumped into the waiting limousines and roared off to the airport. We needed to arrive at the television station in time for the 10:00 P.M. newscast. Zipping out onto the runway, the vehicles screeched to a stop. We scrambled out, thanked the Oklahoma Highway Patrol escort, and ran up the stairs into the waiting aircraft. Unfortunately, in our haste, we had left the pilot at the Terrible Twister show location.

The pilots of our aircraft grew accustomed to unusual happenings. On a flight to southern Oklahoma, one of the KWTV marketing personnel became airsick. She stood up, trying to keep the situation under control. It was hopeless. With a powerful blast, the contents of her stomach hurled through the air and hit the pilot in the back of the head and neck. It was just awful. The pilot instantly arched his spine and lifted his chin upward, but other than that, he didn't move. In fact, after we landed, he didn't want to move at all.

At Hobart, in the far southwestern section of Oklahoma, we were preparing outside the building for a 6:00 P.M. live satellite television feed back to Oklahoma City. I stood there, microphone in hand. The seconds ticked down. I noticed a woman walking toward us. She was moving a white cane back and forth. She was, apparently, visually impaired. Her dog, a furry ball of teeth, was emitting a continuous yipping sound while running from side to side to the full length of its leash.

"Thirty seconds," the camera operator warned.

"Jerry," I called, "that lady is going to walk right into our live shot."

Dalrymple approached the dainty little woman. "Madam, you're about to walk into our live shot," he quietly informed her.

"But I don't want to be in your live shot!" she loudly proclaimed.

"Standby, Gary, five seconds," the camera operator stated, and then "Go," he commanded.

Just as I opened my mouth to speak, Dalrymple reached over and touched the lady on the shoulder to guide her away from the camera. The fuzz ball of a dog detonated. In a rage of fury, it sank its teeth into Dalrymple's leg. I was now on the air. Three feet from me was my boss, his left arm around the little lady and a "killer" dog hanging on for dear life as he tried to shake it loose.

I drove alone to one Terrible Twister show. My destination was Chandler, a vibrant small town just northeast of Oklahoma City. Once there, I became confused about the directions to the show location. After anxiously driving down street after street, I saw my objective. To my dismay only six cars were parked in front of the school auditorium.

It took all of my will power to get out of the car. As I stood at the entrance to the auditorium, my legs felt like they were made of lead. On stage was a large screen and microphone. In the seats, I counted nine people. This was the show that we all feared. No one was there.

I walked in, slipping quietly into one of the many empty seats. It was five minutes before show time. I was unable to make myself walk down the aisle and go back stage. The lights dimmed. A person I didn't recognize walked on stage.

"I would like to thank all of you for coming to our play tonight," the voice said. A fantastic rush of relief swept my body. I was in the right town but the wrong building. Running for my car, I asked a latecomer specific directions to the Terrible Twister show location. It was two miles west and a little south.

Making the final turn, I could see the building. I could also see a figure in the middle of the road. It was Dalrymple. "England, this is the last time you will ever go anywhere by yourself," he announced. I was delighted to find four thousand people inside the building.

The Terrible Twister show is a good tool for reaching large groups of people, but smaller talks at elementary schools are just as important. At an early age, young people can be reached and influenced in a positive direction. Humorous encounters, of course, abound.

While visiting one kindergarten class, a little girl timidly asked, "Mr. England, how long have you been on television?"

When I responded with, "Over twenty years," another little girl instantly clutched her chest and shouted, "Oh, my God!" She obviously couldn't imagine anyone even being alive for twenty years.

Talking with young students at school is usually well controlled and very rewarding. Student tours at KWTV are normally the same. On rare occasions though, a group fails to behave.

After one uncontrollable group toured the KWTV facilities, I received a letter from one of the children's mothers. She had learned of her son's conduct from the adults who had accompanied the band of little rascals.

Among other statements, she said, "I was horrified to learn of my son's behavior." I believe her comment pretty well sums it up.

After the busy winters and wild springs come the relatively peaceful summers. In fact, most summers are so quiet with respect to weather that it's a struggle to keep the weathercasts interesting. During the summer of 1983, I made an attempt to liven things up. Once a week, it was decided, I would invite a guest weatherperson to do the on-air show.

The day before the first guest was to appear, I walked up to my neighbor Larry Cook. "Hey, Larry," I said, "how would you like to come out to the station tomorrow and do my weather show?" He agreed. A few hours later, he claimed he had misunderstood the question, but it was too late. We were both committed.

Larry is short, has receding hair, and tends to become nervous in an instant. On a 10:00 P.M. show, the anchors introduced me, then Larry. With cameras capturing, live, every movement and statement we made, I led Larry to the weather set. "It's yours, Larry. Go to it," I instructed him.

Larry, facing the camera, began speaking in a trembling voice. His shifting eyes looked everywhere except at the camera. At one point, he stopped, looking helplessly toward me. "It's yours, friend," I said. Beads of sweat literally popped up on his forehead. His look was that of a man facing an oncoming disaster with no place to run.

Finally, he finished four minutes of agony. Putting my hand on his back, I congratulated him. He was hot and damp to the touch. He slowly turned to me and with a vacant stare, acknowledged my remarks. His later statements to me are best left unrepeated.

Audience reaction to the guest weather person was excellent. The decision was quickly made to choose another untried person for the following week.

"Phil, this is Gary."

"Hi, Gary," my brother responded, "I saw Larry on the air. Man, was that wild."

"How would you like to do it next week?" I asked. There was a very long pause. "Phil, are you still there?" I knew he was—I could hear him breathing.

"Well, I guess I would," Phil mumbled. It was obvious that he was very unsure about committing to such a crazy offer, but he agreed.

Soon, Phil's day in the sun arrived. He was very uncomfortable as we positioned him for the show. "Sixty seconds," shouted the camera operator. I looked at Phil. He was standing very rigid, tilted several degrees off vertical. His eyes were wide open and were not blinking. The terror of being on television for the first time was engulfing him.

"Are you all right?" I inquired. There was no answer, just a blank expression. "Phil! Look at me!" I demanded. There was no response except for several deep, shaky breaths.

"Ten seconds." Phil looked at me and swallowed. His cotton dry mouth and throat made a loud snapping sound. His passing out on the 10:00 P.M. weather was well within the realm of possibility.

The people of Oklahoma saw, that night, a unique show. Phil, with the appearance of a person on a runaway horse, put on a fascinating program. His facial expressions were priceless. They ranged from fear to hopelessness. It was a performance worthy of "Saturday Night Live."

The following week the guest weatherperson concept collapsed. We were having great success with the idea, so top management became involved. They selected the third participant. Their selection, a cute young lady about thirteen years old, was also very intelligent.

In preparation for her show, I went over the weather presentation with her. She practiced twice, memorized the lines, and went on the air. She performed flawlessly—she really was excellent. What had been amusing was to see people who were not great their first time out. We felt the entertainment value was gone, so we terminated the program.

The Traveling Weather Show   155

Chapter 15

Death and Survival

In early 1984, the next news director arrived. My eleventh boss was young, exceptionally bright, creative and understood the television news business. He knew how to win.

The spring of 1984 in Oklahoma was a rough one. Dealing with another news director and a tornado count that would climb to fifty was a tough combination. I didn't follow closely the changing face of the newsroom, but in the field, tornadoes injured 224 people and killed thirteen. Only one of the fatalities was in our broadcast area.

During the late evening on April 26, 1984, a thunderstorm intensified rapidly just southwest of Chickasha, forty miles southwest of Oklahoma City. The storm quickly became a supercell. Our Doppler showed a mesocyclone signature. Looking at our conventional radar, I could see more ingredients of destruction. The tallest point of the storm was directly over the bounded weak-echo region. There was an overhang on the south side of the storm and a hook echo was clearly defined.

With the storm moving directly over Chickasha, tornado activity was a distinct possibility. The tornado siren sounded the alarm. Writing in the *Chickasha Star*, Fran (McDonald) Snodgrass portrayed well the events and emotions of that night.

The awesome sound of the siren from the alert tower rang in our ears as we attempted to shed our sleepwear. Panic stricken, we felt the urge to be dressed before we took shelter.

Pandemonium had struck our household. Everyone tugged on shirts and other items of clothing, some inside out. Coats and shoes mysteriously vanished or were unseen in our panic.

We rushed out onto the front lawn and stopped. Where should we go?

"The sixteenth street fire station," screamed our daughter.

"No! The neighbor's cellar," yelled another.

The strong wind blew at our backs and filled our mouths and eyes with a gritty substance. The threatening clouds swirled overhead. Lightning flashed and thunder rumbled, as the spring storm bore down on Chickasha early in the night of April 26.

The shrill siren blew on, endlessly it seemed. Cars rushed past, many barely screeching to a stop at the corner. Everyone seemed to be heading in a different direction.

"They say not to get into your car," someone yelled through the noise, and began pushing our little group across the street.

Our back wooden fence fell over, hitting the ground with a sickening thud.

We were met in the neighbor's yard by another couple from down the street. The woman held her large cocker spaniel tightly in her arms. Together we made our way to the cellar beneath the patio, and were out of harm's way.

We stood in the sheltered area and watched as the steady stream of cars sped by. A police cruiser drove up and down the streets with lights flashing and sirens blowing, alerting residents of possible impending danger.

We shivered as hail thundered to the ground and we wondered if the odds had caught up with us.

Fran Snodgras and the residents of Chickasha were fortunate. The tornado did not touch down. Others though, would not survive the night.

I watched the harbinger of death track toward the northeast for nearly 115 miles. Passing just south and east of Oklahoma City, the raging monster maintained its deadly characteristics. No tornadoes, however, were being reported—just high winds and hail.

The supercell should have been producing tornadoes. Within its turbulent structure were all the necessary ingredients. It is still a mystery why some storms of this type produce no tornadoes and others yield one or many twisters.

At about 11:00 P.M., the storm moved out of our broadcast area and into Tulsa's area of responsibility. At 11:33 P.M., with no warning in effect, a one thousand yard-wide tornado plowed into the eastern Oklahoma town of Morris. Eight people in the area were killed, and ninety-five were injured. The twister destroyed twenty-eight square blocks of the small town.

The following day, an injured Morris police officer told of his car suddenly lifting into the air. He and the car flew down main street, smashed into the front of a building, bounced across the street, and slammed into a hardware store—all while the car was off the ground.

With a collective sigh of relief, we welcomed the end of spring and the arrival of summer.

High-tech advances continued throughout the world of television weather. Fall 1984 brought to KWTV a significant advance in severe weather detection: a second generation Doppler radar replaced our historic, first Doppler.

Enterprise Electronics installed our fast fourier transform (FFT) Doppler radar system. It was magnificent. Its tornadic pattern recognition capabilities allowed rapid identification of dangerous storms. The wind velocity spectrum could be examined in much smaller (750 foot) squares. It was a dream come true.

Following quickly behind the FFT was LiveLine 5. This high-resolution computer system generated graphics, images of surface weather maps, and five-day forecasts. Satellite downlinks pumped high-quality, color cloud movies into the computer, which in turn were broadcast to the public.

Also in 1984, telephone data lines were replaced by satellite transmissions. The adaptation of personal computers to meteorology was gaining in popularity. The high-tech movement was picking up speed and would soon become an avalanche.

Computer model forecasts of surface and upper air conditions evolved rapidly. The NWS was soon able to provide much more forecast data. Barotropic, baroclinic, primitive equation, and the limited fine mesh atmospheric computer models became rapidly available. All were designed to project weather conditions in advance. Meteorology quickly became a business of trying to determine which computer model was the closest to correct on a given day.

Alphanumeric computer model output statistics also poured forth. With each new year and new product, it felt more like we were drowning in a sea of information.

Not to be outdone, in early 1985 the KWTV news department purchased Oklahoma's first satellite truck. A miniaturized, mobile television broadcast facility on wheels, it traveled the highways of Oklahoma and surrounding states. Its huge satellite dish, aimed at a target thousands of miles from earth, provided live coverage of news and weather events.

Covering severe weather with the satellite truck was a little dangerous. Newstar 9, as it was known, required a large area and considerable time to turn around. Further, its large, exposed dish made driving in strong winds a definite hazard. Behind the scenes, it became known as the "death wagon."

The engineers that drove Newstar 9 acquired scant severe weather experience during 1985. The entire year in Oklahoma saw just thirty-six tornadoes and thirty-five injuries. It was a calm and welcome twelve months.

The year 1986 was to be another year of below-average tornado production in Oklahoma, but it brought near disaster to the Oklahoma City metropolitam area. Seven tornadoes touched down in Oklahoma County. One was a powerhouse with winds in excess of 160 MPH.

"Alan, ten seconds!" shouted the camera operator. Alan Mitchell, staff meteorologist with me at KWTV, quickly turned to face the camera. Conversation in the forecast center ceased. The rapid whirl of computer-driven printers seemed excessively loud. Ringing phones pierced the air of the dimly lit room. The news radio crackled with reports from the field. A tornado warning continued in effect for sections of far southwestern Oklahoma. Severe thunderstorms were pounding an area from near Kingfisher to Lawton. Also, moments before, a quick, fifty-mile radar scan detected a rapidly developing thunderstorm southwest of Oklahoma City. An air of urgency permeated the forecast center.

Mitchell went on the air, cool and professional. I was sitting behind him at the radar console with quick access to the weather hotlines. Dalrymple, by then vice-president of KWTV and a long-time player in Oklahoma severe weather, was positioned at the station phones. Mike Sims, KWTV assignment editor, stood in the background, microphone in hand, issuing instructions to reporters and photographers in the field, directing them into the path of the approaching storm. As Mitchell moved through his weather show, a continuing stream of personnel surged in and out of the forecast center. To the outside observer, it might have looked like total confusion; but really it was just the opposite, a group of professionals working smoothly and effectively under extreme pressure. The date was May 8, 1986.

The Doppler radar probed the chaotic atmosphere in search of signs of cyclonic shear, sometimes the precursor to tornadic activity. The intensifying storm was moving into southwestern Oklahoma County. On the radar I saw rotating winds at twenty thousand feet. "Azimuth 238.7 degrees, range 26.1 kilometers!" I called to Mitchell. He repeated the information to me as he entered the data into our computer, which then displayed the rotation center location. "This may be the big one," I said to myself. "Increasing rotation and over fifty decibels," I muttered, noting the storm's growing strength.

Suddenly, I was jolted back to the stark reality of the situation, keenly aware that every phone was ringing. "Gary, large hail being reported by a lady along far Northwest Twenty-third!" yelled Dalrymple. In rapid fire sequence, the incoming calls confirmed the severity of the storm. Golfball-sized hail was reported on Council Road, Rockwell Avenue, and MacArthur Boulevard. The storm was exploding, and activity in the forecast center was at a wild pace.

With a slam, I engaged the intercom buzzer. "Yes?" came the reply, from Mary Millen the director.

"Mary, priority-one update, have to go now!"

"Standby!" she responded. Dalrymple was up and in a flash threw open the door to the forecast center. The camera was swinging into place as the lights snapped on.

"Go!" Mary shouted. Without hesitation Mitchell was back on the air, interrupting John Snyder's sportscast with an update on the large hail and serious threat of this dangerous storm.

The pace quickened even more in the forecast center. The update was completed and the lights turned off. Sims was directing news cars to the reported hail locations, while handling a flood of information from the field. "Baseball-sized hail on Northwest Thirty-sixth in Bethany!" someone screamed out through the din of ringing phones and news radio traffic. In a matter of seconds, Mitchell was on the air, advising viewers again. It was now 6:28 P.M. The news ended. We paused for a few brief moments, hoping for the impossible—that it would all stop right now.

As I stared at the Doppler radar, I knew it was not going to stop. In fact, the winds were rotating stronger than ever. "Alan, look at the winds in this thing! Look at this thing! I don't believe it!" I blurted out. Again someone shouted, "Baseball-sized hail just south of Lake Hefner!" Sims then reported rotation visible by one of our news crews north of the Oklahmoa City Fairgrounds. The flow of information became a torrent, nearly overwhelming. The noise in the forecast center was a deafening roar.

Dick Dutton, a former University of Oklahoma linebacker, took charge of the control room. The studio camera operators, Dave Harper and A. J. Johnson, were in position and ready. Bob Lehr was setting up a camera behind the station, pointing west. Rich Kriegel was positioning himself on North Santa Fe Avenue in Oklahoma City. Mike Simon was frantically scrambling north along Interstate 35 to set up at Edmond Road and shoot west. Roberto Prado was moving north on the Broadway Extension toward Edmond—the town that joins Oklahoma City on the north.

We interrupted "Wheel of Fortune" with updates at 6:32, 6:36, and 6:41 P.M. Doppler radar continued to show increasing rotation. A definite mesocyclone had developed. It was decision time. I felt a lump in my throat—and it wasn't from the laryngitis I had been suffering.

There was a severe thunderstorm with large hail and mesocyclone in a densely populated area. The numbers kept flashing through my mind. Approximately half of the thunderstorms with mesocyclones produce tornadoes, usually within twenty to thirty minutes. Rotation had been reported and confirmed by one of our units in the field, and at 6:38 P.M., a viewer reported a funnel cloud north of Northwest Forty-eighth Street.

It was 6:43 P.M. The intercom buzzer shattered the brief calm in the control room. "Yes, Gary," Dutton replied.

"Priority one, tornado warning, take pointer over Doppler radar. I am ready!" I shouted.

"Standby the beeps, standby Mary. Are you ready, Gary?" came Dutton's reply.

"I am ready," I responded

"We're in the beeps, Gary. Cut it and go!" ordered Dutton. I interrupted Vanna White and the Wheel for one minute and thirty-six seconds, advising the viewers of a danger area west of Eastern Avenue and north of Northwest Thirty-sixth Street.

The news director inquired as to the threat of the storm. "No hit reported yet, but it is very serious," I replied. The news director, a master organizer, was in the process of bringing the full resources of the newsroom to bear on the about-to-happen story. Standing there, he stared at me, expressionless. Multicolor reflections from the numerous radar and computer screens danced across his face, creating a ghostly image. I glanced back at the Doppler and then turned quickly to him again, but he was gone. For a fleeting moment, I wondered if we had really talked.

Kim Dalrymple, Jerry's wife, called from an Oklahoma City community called the Village to report heavy rain, then large hail, and then the sound of strong winds passing overhead after the hail stopped. The report gave me a chill. She was describing a supercell tornado sequence!

At 6:55 P.M. I interrupted again with a storm update. The monster was now moving up over North May Avenue and North Pennsylvania Avenue. Funnel clouds continued to be reported along with hail and blinding rain. The Doppler was displaying a fantastic presentation with mesocyclone winds exceeding 60 MPH. The storm was strong and potentially deadly.

Bob Lehr, the camera operator positioned behind the KWTV building, was continuing to videotape the storm which was 4.5 miles west-northwest of our location. At my request, Mitchell joined him to make an observation. At 7:01 P.M., I broke into "Simon and Simon" with the latest information. I defined the danger area as north of Wilshire Boulevard and west of Eastern Avenue and north into Edmond, but there was still no tornado. It occurred to me that we were hitting the warnings so hard that no one would ever believe us again if a tornado didn't touch down.

The flow of information and noise continued unabated. I could feel my pulse pounding in my temples. Suddenly, standing in the door of the forecast center was the normally unflappable Alan Mitchell. Wide-eyed and pointing toward the back door, he said, "You'd better come and look at this, Gary!"

Bursting out the west door, I was astonished to see a large, violently rotating wall cloud exhibiting powerful updraft motion with a funnel cloud trailing from the south end. It appeared to be along Britton Road and near Pennsylvania Avenue. I felt like my heart was in my throat. It was obvious that something terrible was going to happen. Running back to the forecast center, I couldn't shake the thought that people were going to die. Shouting instructions for an update, I checked the Doppler. The mesocyclone winds were now over 80 MPH. It was going to happen, and there was nothing I could do about it! The circulation was moving up over Northwest 122nd Street, west of Pennsylvania Avenue.

Forcing myself to be as calm as I could, I matter-of-factly explained tornado precautions again at 7:06 P.M., pointing out that a tornado could touch down at any time and that residents of Edmond should be especially alert. At this point, I was having difficulty keeping my true emotions from bursting forth. A cool head had to prevail. I kept the warning area north of Wilshire Boulevard and as far east as Eastern Avenue because strong tornadoes sometimes turn to the right.

Sims was continuing to monitor our units in the field and had them in position. "Five-five-eight, this is car eleven," crackled a radio call from Roberto Prado.

"Go ahead, Roberto," instructed Sims.

"What's the movement on the storm?" Prado asked.

"This is five-five-eight. Movement is northeast at 30 MPH, and what is your location, car eleven?" Unbelievably, Prado was near Fifteenth Street and Santa Fe Avenue in Edmond, right in the path of the storm! Mike Simon was taking video from Interstate 35 and Edmond Road. Rich Kreigel was parked on North Santa Fe Avenue in Oklahoma City, also shooting video.

At 7:12 P.M. I updated again, showing live pictures from behind the station. As we watched the real-time movement of the wall cloud, I again instructed the viewers on tornado precautions and defined the danger area as north of Hefner Road, into Edmond. The entire time I was giving the update, I was thinking, "This is it! I can feel it!" The update ended at 7:13:20 P.M. Exactly thirty seconds later, we suffered a power bump. It was finally happening—the tornado was sweeping across the Glen Eagle addition in far northern Oklahoma City. Screaming toward the northeast, it ripped through the powerlines along 150th Street between Western and Santa Fe!

The tornado funnel was visible for only a brief moment. Both Kriegel and Lehr, however, recorded the tornado touchdown and associated powerline flash. Pandemonium broke out in the forecast center as tornado reports poured in. The tornado was moving across a mostly open area toward the Fairfield South addition, intensifying with ground-level winds approaching 140 MPH. We had a potential killer on the loose! Off-duty news personnel were being called in while engineers ran for their remote broadcast vehicles.

At that instant, Dave and Carol Harris were standing in their yard in Fairfield South, looking toward the south. Dave heard the warnings, but he didn't see a classic funnel. Shocked, he did see the top floor of a two-story home in the Glen Eagle addition, one mile away, lift into the air and disintegrate. Dave, unable to turn away from the approaching storm, saw what appeared to be a cone-shaped cloud. It extended from the thunderstorm base toward the ground. The storm was moving directly toward them. Suddenly, debris spewed up from the open field, boiling and churning with violent motion! The tornado seemed alive, pulsing and breathing as it rapidly moved closer.

Carol was clutching their fifteen-month-old daughter, Danelle, when she and Dave saw a large tree beginning to twist. Parts of the tree broke off and streamed upward and outward. The tornado was seventy yards away! "Carol! Get in the house! Get in the utility room! Run! Now!" shouted Dave. Running for her life, Carol quickly took shelter in their utility room, which was centrally located. Dave was still watching the tornado. Panic gripped Dave as he saw part of David and Gina Dabney's home break up, just three lots south of where he was standing.

Seconds now made the difference between life and death. The roar of the tornado was rapidly growing louder. Dave, who had seen the damage created by the multiple-fatality tornado in Morris, Oklahoma, correctly assumed that he and his family were in mortal danger. Jerking open the door, he stepped into the utility room. Carol and Danelle were in a small knot on the floor. "We are going to die!" Dave said. Dropping quickly to his knees, Dave sheltered his loved ones with his body.

Danelle was uttering a continuous, terrified scream. The roar of the wind defied description. They felt like their bodies were going to explode. Dave and Carol felt sheer terror as they heard the house breaking apart. They clutched each other and Danelle, certain their lives were about to end. In ten seconds, it was over. The only section of the house without severe or total damage was the utility room where they had sought shelter. They had survived, unharmed.

Residents of western Edmond were scrambling for shelter as the tornado bored through Fairfield South with winds of approximately 160 MPH. It was 7:15 P.M., and the air was a savage mix of boards, bricks, and even automobiles.

"Eleven to five-five-eight, tornado is on the ground just south of Fifteenth Street! Debris is hitting my vehicle! I'm getting out of here!" came a frantic call from Prado. The twister slashed across Southwest Fifteenth, taking powerlines with it. Mike Simon, shooting video from his position, recorded the powerline flash and the tornado itself.

"Two to five-five-eight, I've got the tornado on video, and from my vantage point it is still on the ground," Simon reported in a surprisingly calm voice. The time was 7:16 P.M. I was on the air again, passing along the tornado location and urging residents to take immediate tornado precautions. And they were doing just that, diving into closets, bathrooms and utility rooms, sheltering their children with their own bodies.

With unbridled fury, the tornado churned through the southeastern sections of the Fairfield addition. The twister smashed directly through the center of one home, destroying the structure and depositing a pickup truck into what had been a child's bedroom. The family survived, huddled in the bathroom. The storm now turned a little toward the right and blasted through the extreme southeastern section of the Trails South addition. Two more powerlines went down in blinding flashes. The storm was moving across Santa Fe Avenue. The winds were still 160 MPH as the tornado turned right and became multiple-vortex while moving through the Copperfield addition.

On the air again at 7:19 P.M., I passed along more reports from our field units, damage information, and repeats of tornado precautions. I glanced at my watch in disbelief. We had been issuing the warning for nearly an hour. When would it ever end? The tornado was now moving almost due east, parallel to Edmond Road, directly toward downtown Edmond. It plowed through the Lamplighter Mobile Home Park. We prepared to go on again.

"Two to five-five-eight, the tornado has lifted," stated Mike Simon. It was 7:22 P.M. and the first tornado had lifted; but another, fortunately small twister had hit the ground just north of the larger one and was tracking rapidly northeastward with minor damage. The severe tornado, however, had disappeared into history.

"Mike, any word on injuries and deaths?" I quietly asked. The question was lost in the uproar that had engulfed the forecast center. I didn't want to hear the answer because I knew people die in tornadoes like that one.

As the evening continued to slip away, we interrupted programming twenty-two more times during the next two hours and twenty minutes. The storms finally started decreasing. It was almost time for the 10:00 P.M. news. "Gary, several minor injuries and no fatalities," Sims blurted out. I was so relieved to hear those words.

As the sun came up on Saturday morning, the extent of the damage became apparent. Debris was everywhere. Smashed cars, with tires pointing skyward, littered the area. Some houses had no walls except for the crumpled remains of centrally located bathrooms and utility rooms. Other homes had one or two outside walls remaining, but the interior walls were gone.

A long, two-by-four board had penetrated an exterior wall. It protruded abnormally from a lone remaining kitchen wall. A can of coffee, apparently undisturbed, was on a cabinet top. A pickup truck, upside down, rested on the flattened remains of a bed.

In all that destruction, it seemed impossible that no one had been killed, and only fifteen injuries, all minor, had been suffered. A total of 41 homes had been destroyed, and another 212 had been damaged. Monetary damages amounted to over six million dollars.

The Edmond tornado was noteworthy, both in its intensity and in its minor physical human toll. Some tornado survivors suffer psychologically for years, but with respect to physical injuries, this tornado experience was nothing short of amazing. It had been an early evening tornado. Therefore, when the tornado struck, many of the residents were home. Why was the casualty count so low?

First, there was at least a twenty-minute warning lead time. Second, Oklahoma residents are extremely familiar with tornado precautions.

Through the years, I have repeated the safety precautions thousands of times. "If you do not have a cellar or basement, go to the lowest level of your home, smallest room in the center part, preferably a closet or bathroom, get down on the floor or in the bathtub and cover your face and head with blankets and pillows."

It seems that most everyone knows what to do when a tornado threatens. In fact, one family believed that the message was understood even by their pet. In the May 5, 1986, edition of the *Cordell Beacon*, Ed Burchfiel wrote,

> Whether you believe it or not, some dogs are pretty doggoned smart. Claude and Thelma Whitener of Cordell have a Pekingese named Pug, who does not have a lot of work to do so he watches quite a lot of television.
>
> It seems that Pug hates thunder. When the weather begins to bang away, Pug runs under a bed or into a closet and hides until the noise stops.
>
> One recent afternoon when the thunder was extremely heavy, Pug disappeared. When the noise ended, Claude and Thelma began to hunt for Pug. Not under a bed. Not in a closet. Finally, they looked in the bathroom. There Pug was in the bathtub. Claude says Pug had heard TV's Gary England advising those in the path of a storm to get into a bathtub.

The Edmond storm contained winds of 160 MPH. Significant, but survivable if you know the safety precautions as well as Pug. If, however, the tornado had produced winds in the range 200 to 300 MPH, a tragic human toll would surely have resulted. Fortunately, tornadoes of that magnitude are rare. They amount to less than 2 percent of the total number that strike the United States each year. They are rare, but account for most of the deaths.

A few weeks after the storm, I was invited to a survivor dinner. There I talked with many of the newest tornado veterans. Their stories were fascinating.

A man and his wife saw the warning, then heard the tornado approaching. Each one grabbed a child. The husband, with his precious human cargo tucked under his arm, dove into the bathroom. With all of his strength, he held on to the toilet and his child. The tornado ripped the toilet out of his arms, but left him and his child on the floor.

I asked him what he remembered just before the tornado hit. He replied, "My wife and other child in the closet frantically pulling clothes off the hangers to protect themselves."

One young mother left her small child at home while shopping at a nearby grocery store. Stepping out of the store, she saw the tornado moving toward her home. Shoving the gas pedal to the floor, her car careened down the street, barely winning the terrifying race. As she leaped from the vehicle, debris was smashing into her home.

She threw open the front door and grabbed her young son. Running for the bathroom, she remembered seeing her cat. The cat showed no concern for the deafening roar of the storm. In seconds the mother was on the bathroom floor, her small child beneath her. The tornado blasted through the home leaving only the bathroom without significant damage. She and the child were safe, but her home, dog, and cat were gone. The young mother stumbled through the debris, tears streaming down her face. She was looking for her dog.

"Laura, don't you have a little black and white dog, slick hair and long tail?" asked a neighbor.

"Yes," she sobbed.

"It's just fine, Laura. It blew though my window."

The cat, blown to some unknown location, also survived. It returned two weeks later, clumps of hair missing and walking with a limp. Laura looked at me and in a matter-of-fact tone said, "You know, they always come back."

During 1986, seven twisters touched down within a fifteen-mile radius of the television station. The Edmond tornado was the largest, but an early morning storm in October was the most unusual. It was born from remnants of a Pacific Ocean tropical storm and swept across western Oklahoma City, inflicting widespread, but minor damage. Again, winter was a welcome relief.

Chapter 16

Official Confrontations

The platform from which all local television programs are launched, the newsroom infrastructure, continued strong but fluid as we moved through 1987. Our sharp, young news director was promoted. My twelfth direct superior was immediately appointed. This news director was a screamer, rendered nearly incapable of normal conversation by the loud and turbulent newsroom environment. Fortunately, tornado occurrence was very low. One source of stress was better than two.

Another news director, however, arrived on the scene within a few months. It was beginning to become a bit confusing—another boss and another set of rules, thirteen news directors in fifteen years.

The year 1988 arrived, and with it came another news director. My fourteenth boss was older, quiet, knowledgeable, and best of all, had a healthy respect for meteorologists and our array of sophisticated equipment.

On March 28, 1988, the news director stood in the weather room and watched a deadly drama unfold. An isolated supercell thunderstorm erupted over Mustang, Oklahoma, just eighteen miles southwest of KWTV.

The radar presentation was awesome. The entire thunderstorm was rotating. A hook echo and a strong mesocyclone were plainly visible. Reports from the field indicated strong wind inflow, hail, and a huge wall cloud. It was late afternoon and many children were at home alone. Other residents were undoubtedly preparing the evening meal and not aware of the impending threat. It was an extremely dangerous, rapidly developing situation.

The track of the storm would take it just north of Will Rogers World Airport, over southern Oklahoma City, Del City, Tinker Air Force Base, and Midwest City. I quickly noted the center of the mesocyclone and the southern edge of the circulation. Tornadoes frequently occur on the periphery of the mesocyclone rotation and not at the center.

"Priority one, tornado warning, take pointer over Doppler radar. I am ready!"

"Priority one, tornado warning. Are you ready?" shouted the director.

"I am ready!"

Nervously, I detailed the location and projected track of the storm. It is a gut-wrenching experience when you know lives are at risk. With the greatest urgency I recommended that residents take immediate tornado precautions, emphasizing that the danger area was as far north as Midwest City and south to Southeast Fifty-ninth Street.

Approximately thirty minutes later a tornado swept across a mobile home park, killing one person and injuring ten others. The location was Southeast Fifty-ninth and Anderson Road. They apparently had not seen or heard our warnings. It was a bad start to the storm season, but fortunately hits by tornadoes during 1988 dropped to a record low of seventeen.

The normal newsroom turbulence continued. Seven months into the year, my fifteenth boss was appointed. His tenure would last until January

1989. In light of the endless parade of news directors, I decided to work more with Jerry Dalrymple, the number-two person at KWTV. I still had to work for the news director, but that was a situation I planned to change.

Dalrymple, one of those people who can look at you with no change in expression for what seems like eternity, had been working to purchase Oklahoma's first privately owned lightning detection system. His tough negotiating style finally succeeded. After twelve months of hammering on the vendor, a low purchase cost was achieved. We installed the system during early 1989. Years later we learned that the vendor compensated for the lower price by using refurbished parts, but it didn't matter. The system worked fine.

Each year, lightening usually kills and injures more people than do tornadoes, floods, or hurricanes. Acquisition of the system, therefore, was a wise move that directly benefited the public.

Lightning and thunder are very unnerving elements of a thunderstorm. As a thunderstorm develops, negatively charged electrons are stripped from their parent atoms by violent motion within the storm. With the loss of an electron, the atom becomes positively charged. A large positive electrical field usually develops in the upper regions of the storm. Correspondingly, a large negative electrical field usually gathers in the lower sections. Most lightning is from cloud to cloud or within the storm.

Even though most remaining lightning originates from the cloud to the ground, a few lightning flashes are initiated from the ground upward. The lower negative field in the storm induces a positive electrical field on the ground that moves with the thunderstorm. The opposite charges gradually grow stronger, and attraction between them increases. Eventually, lightning shoots down in small crooked spurts, called a stepped leader. It is looking for the path of least resistance to the ground. When the stepped leader is less than a hundred feet above ground, an upward moving positive leader leaps up from the ground toward the negative stepped leader. On contact between the two opposites, a return lightning stroke rips skyward from the ground along the channel produced by the stepped leader. This is the lightning flash, commonly called a lightning strike. At this point, several subsequent lightning strokes move up and down the channel. Each lightning flash takes place within a few millionths of a second.

Thunder results from the superheating of the air around the lightning bolt. The air instantly heats to about five times the temperature of the sun's surface, near fifty thousand degrees Fahrenheit. The sudden increase in temperature causes rapid expansion of the air. That expansion creates a shock wave known as thunder.

After acquisition of our lightning detection system, we learned thunderstorms with mesocyclones frequently produced considerable positive lightning flashes rather than the normal negative flashes. Also, during intensification of a mesocyclone, lightning activity rapidly decreased. When mesocyclones reach peak speeds, at which point a tornado is sometimes produced, lighting increased again. Real-time lightning detection became an immediate asset in our severe storm evaluations.

Even though there were few tornadoes, 1989 turned out to be a very eventful year. Another news director was soon appointed, my sixteenth boss. To me, they were even beginning to look alike. And the spring was marked by more confrontations with the NWS.

On March 27, 1989, we were doing battle with scattered severe thunderstorms. At 7:45 P.M., the meteorologist working with me called the NWS. We doubted the validity of an official warning issued at 7:41 P.M. In a follow-up report on the matter, our meteorologist wrote, "I spoke with an NWS employee, and he was very hostile to the fact that we would dare question the warning. He then said the NWS had been watching KWTV all evening and were getting tired of hearing negative comments directed toward the NWS."

The NWS employee was correct on the negative comments, but my dad had instructed me many times, "Son, always tell it like you see it." That's what I do and it makes some people very unhappy.

At 9:38 P.M., my chief assistant called the NWS again. He asked why a warning had just been issued for an area that contained no thunderstorm. In his written report he noted, "At 10:40 P.M., after reviewing the NWS data, I called again and informed them even their own RADAP (Radar Data Processor) data showed their warning to be in error." Again, his phone call was not appreciated.

RADAP was NWS hard-copy radar-derived data. It could be accessed via a computer modem, if one could enter the correct password. On March 29, we attempted to use the RADAP system. Our password no longer worked. A telephone call from the president of KWTV to the NWS elicited no response.

I called a friend at the NWS on March 30. He told me that he "was directed to delete the KWTV password on the RADAP system." He further said that he didn't agree with the action and couldn't understand why the NWS wouldn't admit that "the warning was bad." He then gave me his password into the RADAP system.

A letter from the NWS arrived on March 31, 1989. "Just a quick note to let you know that your new RADAP password has been changed to: Z044P. I hope this change in passwords will be easy to incorporate into your operations." The off-hand tone to the correspondence just served to enhance my anger. I viewed the NWS actions as an abuse of power.

In an April 10, 1989, letter to the NWS I wrote, "KWTV management has asked me to obtain from you a written explanation of why and by what authority the KWTV password and/or account with the NWS RADAP computer was deleted or otherwise rendered inoperative during part of the week of March 27, 1989." No reply was ever received.

During the fall of 1989, the festering relationship between the NWS and KWTV erupted publicly. An official storm spotter, during a raging late-night storm, sighted what he thought was a tornado. The sighting was close to Clinton, Oklahoma. If the information was accurate, lives were in jeopardy. There was little time to waste, and no time to consult with other sources. The KWTV meteorologist on duty made a split-second decision to broadcast the possible sighting and potential threat. A survey the next day, however, could find no tornado damage in the open field where the twister had been spotted—but then again, there was nothing there to be damaged.

Subsequently, a newspaper story about the storm quoted a NWS employee who called the sighting "bogus." In response, on September 4 the official in charge of civil defense at Clinton fired off a letter to the NWS, with a copy to KWTV. It read, in part, as follows:

> I cannot understand why the National Weather Service would want to try to make us look like such fools in the article. The only thing that I can figure is that you people and Channel 9 have some kind of an ongoing feud. The reason that I use Channel 9 is that they are always there when it counts and can answer questions quickly. They also take the time to call us day or night when bad weather threatens. They are very accurate. The last two times the National Weather Service has called us, it was approximately fifteen to twenty minutes after the storm had passed, and our spotters had already been called in. However, I didn't see fit to write a smear article to any papers.
>
> On this particular night two other firemen and I watched the funnel cloud for several minutes as it neared town. If you didn't see it on your equipment, this only tells me that you need to be spending more time tuning your equipment and less time insulting people who are putting their lives on the line to protect their city.

I assumed the episode was over. The television world, however, is full of surprises. Early on the morning of September 12, my phone rang. It was Dalrymple.

"We're taking a drive west this morning. I'll pick you up at ten."

"Right," I weakly responded.

I intently disliked leaving the house early. The normal KWTV work schedule for a prime-time television meteorologist is at least ten hours a day. I considered taking a 140-mile car ride before work an unnecessary perturbation, but when the boss calls, you go.

On the road, Dalrymple explained our mission. By Clinton's invitation, we were going to videotape a meeting between NWS employees and members of the Clinton Fire Department. I thought this was less than a good idea. The weather I could do. This "Sixty Minutes" routine made me very nervous.

With our video rolling, the NWS group came through the door. They were more than a bit surprised. It was, an awkward situation. Initial finger pointing, angry comments, and loud voices eventually gave way to civil discussion. The offended firemen wanted a public apology in the newspaper. It was promised but never delivered.

We ran the bizarre story on the evening news. The reporter, in closing, noted that a member of the NWS group was leaving government service. It was a fair and factual report. Again, I thought the episode was over, but I was wrong.

A local newspaper reporter jumped in to support the NWS employee, stating that the government servant was just taking a sabbatical. Next, a few unfriendly letters to the editor appeared in the Norman, Oklahoma, newspaper and in the University of Oklahoma campus newspaper. Both KWTV and I were the focal points of the criticism. It seemed to me that our attackers were unknowingly doing the same thing they were accusing KWTV of doing—twisting the facts. They were not at the meeting, but nevertheless quoted me. One person who wasn't at the Clinton meeting, for example, wrote: "There [Gary] was feeding the flames and letting the firemen make derogatory statements against the NWS." It has been my experience that firemen are strong, independent individuals. You don't "let" them do anything. They say and do as they want.

The episode was not pleasant. I had not been involved in the tornado warning. It wasn't my idea to video the meeting. I had no control over what the firemen said. I gave no input to the reporter on how to put the story together. But in television, if it's your face on the tube, you take the heat. And you smile while doing so. It goes with the territory.

Chapter 17

Weather Wars

The year 1990 saw the beginning of weather wars. In a mad rush for better television ratings, an extreme escalation in television coverage exploded across Oklahoma. Severe weather reporting and storm tracking grew from ridiculous to irresponsible. It went beyond the limits of sanity and good judgment. I got my first hint of this during an episode of moderate thunderstorms.

A field broadcast provided a storm spotter's account of a tornado on the ground with "trees blowing away and power lines going down." I saw no such indication on radar. During the 10:00 P.M. news, a competing weather person showed a dark, fuzzy video and pointed out the "tornado." I still didn't see it and said so on the air. Subsequent official damage surveys turned up no downed trees or power lines.

A spotter's live audio report of a tornado in southwestern Oklahoma blowing away a barn soon crossed the airwaves. The blow-by-blow description of the destruction made for great television. Again I looked at the radar. There was no indication of a tornado.

The wild babbling of the spotter continued. Within minutes, one of our television field crews called. They were parked within fifty yards of the person giving the dubious report. Our crew reported no tornado and no barn blowing away. Again, official damage surveys could find no damage. It was obvious that storm spotters were creating their own storms, or perhaps were very poorly trained.

The KWTV crews, trained early each spring, are a group of skilled professionals. They are the best in the business. Storm recognition and a calm reporting style are covered in our training, but the main thrust is on safety. All the training in the world, however, could not prepare one of our teams for what happened on March 13, 1990.

Photographer Bill Merickel prepared to leave the station on a storm chase. His partner was the seasoned Dave Balut. Slim, short, and unshakable, Balut helped Merickel load the van for the storm intercept.

Under clear blue skies they left the station, commenting that it didn't seem like a tornado day. Quickly they slipped into the sparse traffic on Interstate 44, a four-lane slab of nearly flat concrete that would take them southwest toward Chickasha. Far to their southwest, bubbling cumulus clouds burst upward, creating huge thunderstorms.

One storm near Lawton, with little delay, assumed supercell status. By this time Merickel and Balut, traveling in tandem with a KWTV microwave truck, were past the last Chickasha turnpike exit. Due to a mix up in instructions, they missed the turn southeast on Highway 19 toward Bradley. Valuable time was lost before the next turnaround was available.

The rotating thunderstorm moved to the southwest of Rush Springs. It was a classic supercell with strong rotation. I felt lives were in jeopardy. Without hesitation, I issued a tornado warning along a line from Rush Springs to Bradley to Criner.

Projections showed that the storm would pass directly over Bradley. I had planned for Merickel and Balut to be near Bradley as the storm approached. Now they would arrive just after it passed.

Large hail and high winds were reported near Rush Springs. Thirty minutes later, the word came in: "Tornado on the ground just northeast of Bradley!" The small twister skipped into McClain County and sliced across Highway 76, five miles north of Lindsay.

Merickel, Balut, and Bill Halford, the microwave truck engineer, had made it to Lindsay. The large turbulent clouds and thick, heavy air seemed a world apart from the conditions when they left KWTV. We directed them north. Merickel, after years of seeking the elusive tornado, videotaped his first.

After videotaping damage along the Highway 76 and interviewing a few people, the crew drove along Alex Road, north of Lindsay. They found more damage and taped more interviews. With darkness rapidly approaching, they returned to the intersection of Highway 76 and Alex Road. It was a good location for transmitting video to KWTV.

Pulling into position, Merickel saw a car parked across the road. He heard the snap of a .22 caliber rifle and said to Balut, "It sounds like there's some good ol' boys over there shooting in the ditch." Switching the interior lights on so Balut could see to write, Merickel kept watching the car.

A rifle barrel was pointing vertically out the car window. The car started to move. It slowly eased through the van headlights. Merickel saw the barrel move from vertical to horizontal. In that instant he knew the rifle was aimed at him. He thought, "They're not going to," and then he heard the sound of glass breaking.

The bullet smashed through the right side of Merickel's chest, punctured his lung, and hit the interior of his rib cage, cracking two ribs. The air in his lung was expelled with such force that he could see his shirt puff out.

Merickel looked at the hole in the window, then put his finger into the hole in his shirt. Blood covered his finger. He turned to the startled Balut. "Yeah, I've been shot," he said.

Feeling no pain, Merickel walked to the microwave truck parked directly behind him. "Those people in that car just shot me," he told Halford. "We gotta go to the hospital."

"Bill's been shot!" The severe weather was over, so most of us were in the newsroom when the message came in. We stood there in stunned silence. Chasing storms suddenly seemed so insignificant.

After a moment, I ran to the forecast center. I found the Lindsay police telephone number and quickly told them what had happened and that the crew would need help finding the hospital. I also told them that Halford was in pursuit of the alleged shooters north of town.

On the outskirts of Lindsay, Merickel's lung collapsed. Every breath was a struggle. His pain was intense. Calling on his Kung Fu training, Bill focused on staying alive.

At about the same time Halford reached a standoff with the alleged shooters. They stopped, guns protruding from their vehicle. Halford quickly threw the truck in reverse and backed wildly down the road to an assumed safe distance. He waited.

"Get me there, Davy, get me there," Merickel whispered to Dave. In Lindsay they met a police car running hot, code three to the north. They thought the police were going after the person who had shot Merickel, but in fact they were looking for Merickel, whom they mistakenly believed was in the microwave truck. The police, however, did find the KWTV truck and arrested the people involved in the incident.

Slumping over on Balut's arm, Merickel blacked out. Balut shook him and shoved him into a sitting position. Merickel regained consciousness, but then they made a wrong turn and could not find the hospital.

Out of nowhere, it seemed, a Lindsay resident flagged them down. "I've been listening to the police scanner. You're going the wrong way. Follow me!" he shouted.

In the emergency room Merickel's clothing was quickly cut away. A doctor held up a long tube and explained that it would have to be inserted into the chest. Intravenoud fluids and medication were started.

Merickel looked at the two IVs. His gaze then turned to the people in the room. Everything started zooming away, getting darker, fading to black. He fixed his eyes on the nurse and said, "You all better do something pretty darn fast because you're about to lose me." At that point, in the doctor's opinion, Merickel died.

Medical technology and skilled people refused to let him go. Merickel resisted his return. He could faintly hear a nurse calling his name, but he mentally fought her insistent attempts to communicate.

With Oklahoma Highway Patrol cars leading the way, Merickel and the medical team took off in an ambulance. During the long ride to Baptist Medical Center in Oklahoma City, he regained consciousness, with the same persistent nurse at his side. The ordeal changed his life. It also gave me a different perspective on what is really important in life.

The weather wars, though, continued unabated. On a stormy Sunday evening, thunderstorms moved toward Oklahoma City. They were heavy, but decreasing in intensity. One young television meteorologist at another station repeatedly interrupted programming to warn of the dire threat to Oklahoma City.

Finally, in a fit of desperation, I interrupted KWTV programming to explain that no severe weather threat existed. This was not a good idea. The program was "Sixty Minutes." Many viewers called KWTV to tell me to stay off the television when there wasn't any threat of severe weather. Angry and frustrated, they felt that enough was enough. I felt the same way, but had used the wrong approach.

On the afternoon of May 26, 1990, I talked with my neighbor about the out-of-control weather coverage. As we chatted, a few small cumulus clouds slowly appeared in the pale blue sky. An outflow boundary created by collapsing thunderstorms to our northeast was moving toward the southwest. I could see it rapidly approaching Oklahoma City. Even though no severe weather was expected, the cool air was moving toward very hot, moist, and unstable air just west of Oklahoma City. I had been looking forward to having two full days off from work. It wasn't going to happen. Excusing myself, I made my way to KWTV.

Thunderstorms soon erupted along the southwest-moving boundary. By 6:00 P.M., the southwestern shove had stopped. A large thunderstorm just west of El Reno developed a significant mesocyclone. A tornado warning for Canadian County was broadcast. Another intense thunderstorm rapidly developed just north of Hinton. Organized circulation in the storm appeared weak.

At 6:50 P.M., the phone rang. "Gary, I live just north of Hinton. I see a funnel cloud to my north. What do you see on radar?" my caller inquired.

"I don't see a tornado on radar, but there could certainly be one there," I responded. This surprised him. Many people think that Doppler radars detect all tornadoes, but they don't.

A second call from the same area soon came in, also reporting a funnel cloud. Since Doppler showed only weak rotation, I commenced a detailed evaluation using our older black-and-white radar. A small, tightly wound hook echo was plainly evident. I called the Hinton Police Department. "We think there may be a tornado just east of your town," I told the dispatcher.

The dispatcher did not hesitate. A patrol car was immediately sent to the east side of Hinton.

Home video sent to KWTV later tells the story quite well. With the camera pointing east, the patrol car comes into view at the same time a large tornado touches down. The tape shows that the police officer flipped on his warning lights and siren, turned the vehicle around, and sped quickly toward police headquarters. For some reason he was unable to communicate by radio with the dispatcher so he had to drive to headquarters to sound the alarm. I immediately issued a tornado warning for the Hinton area.

The twister came down two miles east of Hinton. It moved slowly toward the southeast, then toward the west and finally toward the south. On the ground for three miles with winds estimated near 200 MPH, the storm destroyed the few structures that were in its path.

At one home, a resident took shelter in her cellar. As the noise of the approaching storm grew louder, she became more terrified. The severe tornado passed slowly over her cellar. The noise was deafening; the invisible air pressure, powerful. In a split second, the cellar door was ripped away. Everything not secured streamed up and out into the spinning cauldron.

The security of the cellar was a blinding torrent of tiny debris. Seconds seemed like hours. The horrific winds tried to claim her as their own. She grabbed the leg of a built-in cedar table. Her body went horizontal. Only her hands maintained contact with the world she knew. Holding on for her life, she fought the tornado as it tried to suck her out the door. Slowly the violence moved away. She fell to the floor, alive but riddled with small slivers of wood. It was two years before she could talk about the experience.

The lives of several people were saved that day, but not by overzealous television reporting from the field, and not by fancy government and commercial Doppler radars. Lives were saved by timely calls from private citizens as well as the quick actions of civil defense and law enforcement agencies.

The media battle on this storm was waged over coverage of the damage. All night long, crews from the competing stations maneuvered to find the best broadcast location. Real-time encounters with tornadoes had to wait for another time.

From the ongoing competition some good did come. One organization, using Sony and Cellular One equipment, managed to send still color pictures via cellular telephone. It was a significant advance in the transmission of visual images from the field.

Not to be outdone, in 1990 we developed a computer program that placed a small state icon in the corner of viewers' television screens during severe weather. It was part of an automatic warning system called First Warning. It is now sold across the country under a different name. Some viewers love it, but others object when it covers up part of their favorite program.

In 1991 we developed a severe storm tracking program dubbed StormTracker. At the same time we decided to call our storm chasers StormTrackers. For the first time ever, the time of arrival of severe weather could be projected for any location and shown to the audience. Public reaction to the StormTracker system was extremely favorable.

Early spring 1991 brought a flurry of tornadoes to the Ada, Oklahoma, area. During one of my warnings, I used StormTracker to project that the tornado would hit north Ada at 4:33 P.M. Later that night, one of our photographers showed the video he had shot. On it was a man who had been in the tornado. "Gary England said the tornado would hit Ada at 4:33 P.M.," he said. Pointing to a demolished structure he continued, "And that's the time the clock stopped in that garage that used to be there." That video was a priceless tool for promoting KWTV weather and enhancing the reputation of the meteorologist.

Early in 1992, KWTV installed a new $400,000 Enterprise Electronics Doppler radar system. Our immediate dilemma was what to name it. For promotional purposes, television stations have names for most all of their equipment. Some television stations choose names or phrases similar to what the competition is using in hopes of confusing the audience and diluting the leader's position.

A viewer letter described the problem. In reference to a competitor she wrote, "I don't appreciate the way he constantly slurs you but has no problem in stealing your ideas and then making it seem like they are his own."

We settled on the name Doppler 9000, after our channel number. It seemed unlikely that our major adversaries would name any future radar they bought Doppler 10,000, as their channel numbers were four and five.

Doppler 9000 was the newest electronic marvel available. After considerable pleading with my superiors, it was decided that we would operate the new Doppler along with our older Doppler. Two real-time Doppler radars plus NEXRAD turned out to be a strong asset in our continuing battles with the raging storms of the southern plains.

Storm Action Video made its debut in 1992. It was a significant advance, developed by Roger Cooper, our main news anchor. He installed a Macintosh computer in a vehicle and another at KWTV. Using video compression, he then transmitted near real-time moving video via cellular telephone. Still pictures had been transmitted before, but never moving video.

The KWTV Colby video system followed closely on the heels of Storm Action Video. It allowed, for the first time in history, the transmission of real-time moving video over cellular telephones. The picture was a bit blurry, but the Colby system represented a notable advance.

Ironically, two years later another television station purchased a similar system, entered it in the Emmy Award competition as new technology and won.

These advances were excellent but nearly offset by the increasingly hysterical approach to severe weather coverage. Fleets of television storm spotting crews roared down the highways. One television station crew careened about Oklahoma in a van that contained weather radar. "Blue sky" reports—broadcasts that contained no significant information—were becoming standard. My favorite of these was a live broadcast from southern Oklahoma. The reporter, standing in a small town, pointed toward the deep blue sky and a few fluffy clouds. "These clouds may soon grow into thunderstorms that may soon produce tornadoes," he wildly stated. They didn't.

On September 2, 1992, however, a few towering clouds did grow into massive thunderstorms. It was like spring in the fall. A strong mesocyclone developed northwest of Purcell. The storm, about thirty miles southwest of KWTV, moved in an easterly direction.

"Priority one, tornado warning, take Doppler 9000. I am ready."

"Are you ready?" came the all-too-familiar response.

"I am ready."

The director, faceless to the public, swiftly moved to action. With fingers that seemed to dance, he quickly engaged color-coded buttons from a multitude of selections.

After issuing a tornado warning, we concentrated on positioning our field crew. It appeared the storm would pass just north of Purcell. Our crew consisted of Bill Halford, engineer; Larry Lockman, photographer; and Val Castor, StormTracker. They were located just south of the projected storm path.

Halford, in his microwave truck, raised the large telescope antenna and established video communications with KWTV. The view of the developing tornado was dramatic. Rotation was well organized and intense. The storm suddenly turned toward the southeast. It was 5:45 P.M.

The new course would take the circulation just west of our field crew. Their position was dangerous. Lightning flashed around the area. A thin rain curtain wrapped itself around the mesocyclone, which was increasing in speed by the moment. It was an unreal view—two thin sheets of rain, appearing as one, moving in opposite directions.

Halford was first to feel the ground electrical charge build. His skin hairs tingled. Due to safety training, he knew lightning was about to strike. Instinctively he grabbed the truck door handle. It was the wrong move to make. At that same moment, a nearby ground flash occurred. A powerful electrical surge pulsed through Halford's body. Lockman, standing nearby, felt the surge too, but not as strongly as Halford. Castor, intently watching the huge circulation just to their west, was not aware of what had happened.

Halford and Lockman were shaken but seemed unharmed, so the crew quickly prepared to head south on Interstate 35. Halford lagged behind to secure the microwave truck for travel, while Lockman and Castor set off, attempting to position themselves to the south of the storm. They didn't make it.

The cellular phone in the station rang. "Gary, this is Val! We're just south of Purcell, doing sixty-five, and debris is passing us! We're in the tornado!"

The tornado, five hundred yards wide and with winds of 150 MPH, swept across the highway, flipping several cars just in front of Lockman and Castor. The twister quickly ripped the top from a stand of trees to their left, then disappeared into the distance.

Halford, Lockman, and Castor spent the next few hours taking video of damage and sending it back to KWTV. In the solid darkness of that fall night, they finally headed back to KWTV. South of Norman, Halford felt very ill. Lockman and Castor took him to the closest hospital. Halford was suffering from the lighting strike. Lockman was also injured, but to a lesser extent. None of us realized that it sometimes takes hours for lightning injuries to manifest themselves.

Halford survived but has not worked since that day. Lockman still chases storms, but is much more aware of the constant danger that a thunderstorm poses. Castor continues to pursue elusive tornadoes anytime and anywhere.

In response to their experience, KWTV immediately instituted even stricter safety guidelines for our staff and StormTrackers. In fall 1992 we also initiated a plan to ensure that we were not involved in unnecessary storm coverage, but in television, it can take the audience a while to figure out who is doing what. Until that time everyone is lumped together in the same ugly pot.

The television tornado frenzy had become the subject of countless conversations, letters, phone calls, and newspaper stories. Television program interruptions had increased in frequency and duration. False sightings of tornadoes, inflated wind speeds, and larger than life hail size reports had become the rule. Even winter storms lost their purity. One reporter, broadcasting live, positioned himself behind a very small snow drift—squatting down, to make the drift appear huge. On another report, a spotter told the audience that slick roads had traffic nearly at a standstill, while behind him vehicles could be seen moving by at a rapid rate of speed.

J. B. Bitter, with the *Enid News and Eagle*, however, understood exactly what was going on. On Sunday, March 21, 1993, she wrote,

> Tornado coverage is the Oklahoma version of tabloid television. I predict spinoff talk shows at any time: Today's topic, bovines that have been sucked up in cyclones.
>
> Tornadoes are a part of Oklahoma's heritage. They have proven a deadly part of our history and technology has no doubt saved lives by providing early warning.
>
> But isn't early warning more valuable when it comes in the form of Gary England cutting into "60 Minutes" to tell you to grab the kids and the dog and head to the cellar?
>
> As opposed to, say, some guy with a microphone in his face and his hair being sucked from atop his head, yelling, "We're tracking the funnel cloud across a wheat field just to our left, Jack." Nobody's going to jump up and run for cover. They're all glued to the TV to see if the weather spotter is going to be woven into a barbed wire wall decoration.

To a certain extent, we at KWTV were initially caught up in the weather frenzy. None of our people intentionally broadcast false reports, but we did too many broadcast interrupts for situations that did not warrant the attention.

It had always been the norm to watch the competition to make sure they were not doing better than we were, but such viewing can also have a negative effect. I felt like we were being sucked into a dark, meaningless pit. Something had to be done. The competition can influence your actions by what they are showing or saying. I called it the "lemming effect," after the masses of rodents that panic and eventually run into the sea, destroying themselves. After two years of escalating and questionable weather coverage, I ordered the weather department not to view our competitors during severe weather. Television sets clicked off in the forecast center. VCRs clicked on. Eventually the tapes captured the peak of irresponsible severe weather broadcasting.

On May 8, 1993, high winds, hail, flooding, and tornadoes were affecting a large part of Oklahoma. One of our competitors came on the air with a bulletin. A television storm spotter, in a panic-stricken voice, described a tornado on the ground in Ryan, Oklahoma, with "houses exploding." The attending television weatherman added his solemn comments to the lengthy drama. Again it was fantastic television, but again it wasn't true.

There were tornadoes in the area, but the town was not hit and no houses exploded. In fact there was no damage at all in the town of Ryan, Oklahoma. That type of weather coverage struck me as similar to shouting fire in a crowded theater.

Residents of Ryan told our reporter of frantic calls from relatives who feared the worst. An Oklahoma City Church of Christ called the denomination's church in Ryan to find out what relief supplies would be needed. Later a representative of the telephone company told me that viewers had placed so many calls to the area that the emergency 911 number was shut down in some towns near Ryan, a tremendous risk in the event of a real emergency. On hearing of the apparent destruction, the local power company prepared to call in extra crews to restore power in the devastated area, a devastated area that didn't exist. Reckless reporting of severe weather obviously has wide-ranging consequences.

The real tornado that day struck Healdton, Oklahoma, some thirty miles northeast of Ryan. A few days after the storm, Bob Humphrey, a Healdton resident, wrote, "The last thing we heard from the television before power went off was your warning that the storm would be in the Healdton area at 7:13 P.M. Today, we had electrical power for the first time since the storm. The last time printed out on the log of our fuel computer was 7:12 P.M., May 8, 1993." He continued, "Thanks again for your efforts and the long hours you and the TV 9 staff give so that others can feel secure living in Tornado Alley." It was obvious that our rejection of the "sky is falling" routine, combined with our use of new technology, was continuing to save lives. His letter helped to confirm that we were on the proper track.

Also on May 8, severe flooding hit Oklahoma City. Fast-running waters swept five persons to their death. In a dramatic, dangerous rescue, Ranger 9 pilot Leroy Tatom plucked two individuals from their stranded vehicle. A letter from Peter Erodes to Leroy Tatom convinced us even more that we were doing what should be done.

Mr. Erodes wrote,

> I was the individual who made the 911 call concerning the two people who were trapped in the van on Hefner Road during the flood. I want to commend you [Leroy] for your courage and quick thinking in saving those people.
>
> While standing there and watching the drama unfold, I became convinced the people were going to drown. Most television viewers do not realize that you risked killing yourself, given the nearby power lines, when you saved them.

It is quite refreshing to know that while the other two local television stations are too busy sensationalizing the news, chasing storms in other states, and irresponsibly making false weather reports, you and Channel 9 are actually contributing to this community.

Thank you once again. Heroes such as you are rare in this day and age.

Yes, Leroy was a hero—an ordinary person who carried out an extraordinary act.

Chapter 18

Full
Circle

Severe weather was minimal during spring 1994, but I still looked forward to the hot, inactive days of summer. My seventeenth and eighteenth news directors had come and gone a few years back. The nineteenth news director was in residence. I, however, no longer worked for the news director. The KWTV weather operation was finally functioning as a separate department, on an equal level with all other departments within the company. It was a good feeling. We had a fair degree of independence, leading-edge technology, and quality meteorologists. Ours was a top-of-the-line operation.

Storms were unusually frequent in July. Days were long. Meteorologists became weary and short-tempered. Storms hammered away, usually during the middle of the night. August was no different.

On August 7, thunderstorms were moving toward the south, east of Oklahoma City. It was early on a Sunday morning. Many people were in bed, some were watching television, and others were preparing for church. Our meteorologist broadcast a warning for Lincoln County a few minutes after 7:00 A.M.

No damage reports came in as the storm moved toward Prague. The thunderstorm was crossing our radar beam at a ninety-degree angle. Any north-to-south winds were invisible. At 7:30 A.M., a powerful, long-lasting downburst engulfed Prague, a small town located in the southeast corner of Lincoln County. It was a terrifying surprise for many residents. Nearly every structure was damaged to some degree by winds that reached nearly 100 MPH.

I soon received a letter from a resident of Prague. She had awakened after our broadcast and assumed we had not issued a warning. She was very angry and told me so in no uncertain terms.

As I read her letter I thought about how fast a television meteorologist can go from good guy to goat. Without a doubt, one storm can do the trick.

An inquiring voice brought me back. "Gary, how do those storms in Kansas look?"

"Still coming south, hard and fast. It's going to be a priority-one day," I replied.

The thunderstorms neared the Oklahoma border. Surging south, the line formed into a shape that looked like an arrow. There was definitely high potential for damaging winds and large hail. It was August 17, 1994. The thought crossed my mind we were living through the summer that "roared."

At 1:18 P.M., as the thunderstorms moved into Oklahoma, I issued a heavy thunderstorm advisory. At 1:47 P.M., moderate-sized hail flew across the sky. Winds of 70 MPH snatched a barn door from its hinges. A few minutes later, a viewer called. She lived on the north shore of the Great Salt Plains Lake. The winds were so strong that the water had moved several feet toward the south, essentially altering the shoreline for a brief period. That was not good news.

The clock showed 2:35 P.M. "Lahoma is blowing away, the hail is gigantic! It's unbelievable!" screamed an unidentified voice over a noisy phone line. The line went dead. A cold shiver captured me. I was already very concerned, and that call left me nearly speechless.

A titanic, long-duration downburst blasted south through approximately eighty square miles, from five miles north of Lahoma and Meno to five miles south of Drummond. Hail larger than baseballs stripped trees of foliage, destroyed automobiles, and took the paint off all north facing structures. In some homes, hail blowing horizontally left gaping holes in walls. Structures and trees in Meno, Lahoma, and Drummond were shredded. Sustained five-minute winds of 78 MPH with gusts to 113 MPH were recorded before the measuring instrument was damaged. It was a horrendous encounter for the people caught in the intense violence.

Program interrupts for warnings were frequent and deadly serious. Our StormTracker field units were positioned to intercept the storms, a dangerous job in this type of situation. Val Castor and Rob Satkus waited south of Kingfisher.

"The storm is moving into Kingfisher with baseball-sized hail. You're directly in the path. Go south quickly," I advised Castor and Satkus. The two graduate meteorologists, seasoned and nearly fearless, decided to maintain their position for a few minutes.

A small tornado touched down southwest of Kingfisher. Huge hail slammed into their vehicle, cracking the windows and creating a moonscape surface on the metal. They scrambled south, taking refuge under an overpass.

The line of thunderstorms managed a few more spurts of large hail and damaging winds. El Reno and Mustang, both located to the west of Oklahoma City, were the recipients.

The storm was like a hurricane with hail. Extreme hail. The windows of a pickup truck caught in the core of the storm were shattered into tiny bits. In the back of the pickup was a gigantic piece of hail. It was estimated to be seven inches long and four inches across. Official damage reports listed the hail as football-shaped.

Hail forms when small pieces of ice are continuously rotated through the updrafts and downdrafts within a thunderstorm. Inside the thunderstorm there is a large zone of supercooled water, water that is below freezing but still liquid. Each time a piece of hail moves through the supercooled water, it acquires a new coating of ice. At a certain point, the hail becomes too heavy for the updraft winds and falls to the ground. Most hail falls straight down. But on August 17, 1994, it moved horizontally on winds of 113 MPH.

On the demise of the storms just to our west, new storms developed in far western Oklahoma, north of Cheyenne. By this time it was late afternoon. Those of us in the forecast center were growing a bit weary. The ever-excited StormTrackers, though, raced west on Interstate 40 to intercept the storms.

High winds and hail pounded through Cheyenne. Elk City suffered widespread damage. West of Elk City, high-velocity winds scattered eighteen-wheelers across the landscape. It was a repeat performance from earlier in the day. Devastating straight-line winds, a small tornado, and large hail were sprinting southward along a front three to eight miles wide.

The few structures in the tiny town of Carter were flattened. Three-foot-diameter trees were sheared off. Power poles were snapped like dry sticks as the boiling mass took aim at Mangum, Oklahoma. The storms were unusual, even by Oklahoma standards.

Mangum, in far southwestern Oklahoma, was soon hit. At 8:04 P.M., winds of 100 MPH and large hail pulverized the town. Power lines swung wildly in the onslaught. Roofs were ripped away. Shingles, boards, and other debris sliced through the air in the horrendous uproar. Leaving the dazed residents of Mangum behind, the thunderstorms moved swiftly south into Texas. The damage path was forty-two miles long.

The day-long siege significantly damaged nearly 1,400 permanent structures. In addition, 128 mobile homes were destroyed or damaged. Hundreds of power poles were downed, and thousands of vehicles were damaged.

I looked at the clock. It was 9:24 P.M. I had been in my chair at the warning position since 1:18 P.M. We had broadcast sixty-eight warnings and updates, not including the weather presentations during the regular 5:00 and 6:00 P.M. newscasts. Our weather bulletins, averaging one every seven minutes, totally disrupted normal television programming. This time, there was not one complaint. But it had been a life-threatening situation.

The August days seemed to tumble by in slow motion. Storm after storm raked the plains of Oklahoma. During idle moments in the darkened forecast center, the angry letter from the Prague resident fluttered through my thoughts. The old man, his mules, and wagon of so long ago were there, too. Faded memories of Harry Volkman shimmered in the background. My mind kept asking the same question, "Why would anyone be in such a business?" It requires total commitment, on one hand, and, on the other, exposes the slightest imperfections for all to see and criticize.

I looked at Doppler 9000. New storms were going up. I felt a surge of excitement, and then I knew. It was the thrill of trying to outsmart wild, raging storms that kept me involved in the turbulent world of tornadoes and television.

It's a rapidly changing world, though. Gone forever are the days of matching wits with Mother Nature on a one-to-one basis. More and more, the television meteorologist is becoming a manager and presenter of data. Technology has become the brilliant assistant that provides a swift analysis, evaluation, and projection of any storm threat.

The electronic marvels, though presently far from perfect, will continue to advance. Romancing the storm will ultimately fade into the past. Information on weather will reach parity among competitors. To survive, the television meteorologist of the future will have to become the viewers' true friend. Traveling full circle back to the warmth, charisma, and concern of Harry Volkman will be essential for success.

Chapter 19

Distant Agony

The spring storm season of 1995 sputtered to a slow start. A few severe thunderstorms erupted. Abnormally cool air, though, kept storm intensity to a minimum. News director number twenty was in residence, ready to cover the damage that was sure to result from Oklahoma's famous weather.

It was an exciting time. I had been asked to play a small part in the Steven Speilberg movie *Twister*. I quickly accepted. Warner Brothers also requested that I provide their crew with a severe storm safety meeting. I didn't have to think twice about that decision either.

On a sunny Wednesday morning, I prepared to broadcast my last radio weather forecast for the day—on KXXY Radio with Dave, Dan, and Nate. I had been working with them every weekday morning for over six years. My broadcast facility was on the second floor of my home.

I waited for the 9:00 A.M. commercial break to end. Suddenly I heard a loud boom. In a fraction of a second my house heaved upward. It creaked and groaned as a ground shock wave ripped by right after the air blast. It was 9:02 A.M., April 19, 1995.

"Mary," I shouted to my wife, "go outside and look! Something terrible has happened.!" I was sure an aircraft had crashed nearby.

With just a few seconds left in the commercial break, I dialed the KXXY Radio hotline. "Nate, this is Gary. We've had a big explosion out here."

"Gary," Nate replied. "I see smoke pouring out of downtown. It doesn't look good." Nate's normally solid voice was not the same.

"My God," I said, "I live ten miles from downtown and it felt like the house was coming down!" But Nate had already hung up the phone and was introducing my weather broadcast.

I gave a quick weather forecast, then we all spoke briefly about the fact that something had happened. Little did we realize the magnitude of the disaster that was beginning to unfold.

I snapped the radio line to the off position, dialed the television station and found all circuits busy. I tried several more times, but the result was the same.

I ran down the stairs and out the front door. I was unable to see smoke or damage. I could hear, though, the distant wail of sirens.

The next few minutes were a blur. It seemed like it took eternity for the television to respond. I kept dialing the television station. All circuits were still busy.

I quickly clicked through the channels. Nothing, nothing, and then on KWTV, I saw our tower camera view from six hundred feet above ground level. It was aimed at downtown Oklahoma City. Two huge debris and smoke plumes were visible. My wife and I stood there in stunned silence. Something ungodly had happened.

In a few moments, Ranger 9 live video scrambled across the television screen. Slowly, it seemed, the picture became more clear. The helocopter approached downtown from the northeast and made a wide counterclockwise circle around the area that was belching black smoke. Visibility was poor.

I couldn't move. The Ranger 9 camera gradually pointed toward the northwest and then the southwest. The view became clear and horribly real. A large part of the Alfred P. Murrah Federal Building was gone, crumpled on the streets below among the burning cars.

"Car bomb," I said to my wife Mary.

Indescribable emotion filled my mind. I knew, without a doubt, that many people had just died. People in the Murrah building who the night before had been part of my television viewing family were gone. I knew, at that moment, many children were now without parents. I didn't realize, though, that many parents had lost their young children.

Across the street, I learned later, employees at the Oklahoma Water Resources Board lay dead and injured. They were wonderful people with whom, only a few months before, I had met, laughed, and talked with as they raised money for the United Way.

All of my working life in television I have labored to educate the public about tornado safety, spending endless hours teaching men, women, and children that they can survive the wrath of a tornado. On February 22, 1995, I met with the Oklahoma City Police Emergency Team to discuss what they should expect to encounter after a major tornado. I'm sure none of us envisioned the devastation that our city would eventually suffer.

Now I stood there, viewing a bloody carnage for which there was no warning, no protection. In a few minutes KWTV gave CBS and CNN permission to broadcast their signal around the world. Millions of people must have felt as I did, helpless.

The television station was flooded by phone calls and television correspondents from across the country. Every person in the station was helping in some way. Inside KWTV as well as at the bomb site, it was a scene of agony and caring that words cannot describe except to say that the KWTV news people were magnificent in their coverage and personal control. They showed the world, with dignity, a dreadful tragedy, but they also showed the birth of countless heroes who acted in the face of great personal danger. Fire fighters, police officers, doctors, nurses, and men and women from all walks of life struggled to rescue the survivors.

To add to the misery, the weather turned bad that night. The KWTV weather staff was requested to provide around-the-clock weather reports to Assistant Fire Chief Jon Hansen and Weather Officer Clint Greenwood. Bathed in bright lights, the television view of the bomb site was surreal. Strong winds buffeted the area while I issued tornado warnings for southern Oklahoma. It was a day and night designed in the depths of hell.

Eventually the death count reached 169 men, women, and children. Nearly 500 persons were injured. Damage to the Murrah building and dozens of others buildings was estimated to be in excess of one hundred million dollars. The numbers are staggering and larger than any tornado is likely to produce.

Most of us will never be quite the same as we were prior to 9:02 A.M. on April 19, 1995. Now when I issue a tornado warning I am more keenly aware that someone could be killed in the storm. And for the first time, I realize that each moment we live is a gift.

Glossary

Advection. Movement of temperature, moisture, etc., from one point to another.

Aerographer's mate. Navy term for an enlisted person whose specialty is weather.

Air mass. A large body of air in which the horizontal temperature and moisture distribution is fairly uniform.

Air pressure. Weight of the atmosphere.

Anchor. The primary presenters of television news, weather, and sports.

Anomaly. A value or event that does not fall within an expected normal range.

Anticyclone. A high pressure area, which in the northern hemisphere has winds rotating in a clockwise direction.

Anticyclonic flow. Winds moving around a central point, in a clockwise direction in the northern hemisphere.

Anvil. Cloud that extends around and downwind from the top of a thunderstorm. Made up of tiny ice crystals.

Attenuation. Loss or intentional diminution, of radar signal power.

Azimuth. Direction of a target (e.g., a storm) from a given center point of a 360° compass. North is at 0° (or 360°); 180° is south.

Bounded weak echo. Radar image that shows an area of weak precipitation returns surrounded by stronger returns. Frequently suggests the existence of circulation.

Chief petty officer. Navy term for a supervisory officer among enlisted persons who have reached management level.

Cloud. A visible mass of tiny water or ice crystals.

Cold front. A zone of transition where colder air is replacing warmer air.

Colorgraphics. Term used to describe weather graphics seen on television.

Condensation. The process by which water vapor is converted to water. Heat is released during this process.

Control room. Area from which the television director and crew control television audio and video transmissions.

Convection. Transfer of heat upward into the atmosphere. Frequently creates thunderstorms.

Cyclone. A low-pressure area, which in the northern hemisphere has winds rotating in a counterclockwise direction.

Cyclonic flow. Winds moving around a central point, in a counterclockwise direction in the northern hemisphere.

Dew point. The temperature to which the air must cool in order for condensation to occur.

Director. Individual responsible for directing live television broadcasts.

Doppler radar. Converts the shift in frequency of moving targets into speed. Measures wind speed and direction of movement of precipitation particles. Also measures precipitation intensity.

Downburst. Potentially damaging winds moving straight down from a thunderstorm and then producing a horizontal zone of influence greater than two miles wide.

Downlink. Describes the reception of satellite television transmissions.

Dryline. A boundary where adjoining air masses exhibit extremely low moisture content on one side and extremely high moisture content on the other side. When accompanied by high air temperatures, thunderstorms may form along and east of the dry line, usually oriented from north to south.

Executive producer. Individual responsible for supervising producers. Also implements and oversees news format and production of newscasts.

Flanking line. A stair-step line of towering cumulus clouds merging with the primary thunderstorm, usually on its trailing southwestern flank.

Front. A zone where air masses of different properties are in contact with each other.

Funnel cloud. A tube or cone-shaped rotating cloud not in contact with the ground.

General manager. Individual usually in charge of the entire television operation.

Hail. Ice produced by thunderstorms, ranging in size from tiny to larger than softballs.

Ham. Amateur radio operator.

High pressure area. Anticyclone. An area where the net movement of air is down and then away from the center. In the northern hemisphere, winds flow clockwise around the center.

Hook echo. A radar precipitation pattern that looks like a hook or the numeral six. Sometimes indicates the presence of a mesocyclone and a tornado.

Hurricane. A tropical low-pressure area that is usually hundreds of miles across. Has sustained winds of 74 MPH or greater.

Inversion. A condition in which the air temperature increases with height and then at a certain point sharply begins to cool.

Jet stream. The core of high-speed winds that can exist at almost any level of the atmosphere.

Lightning. Visible electrical discharge within thunderstorms—cloud to cloud, cloud to ground, or ground to cloud.

Low-level jet stream. Band of strong winds approximately three thousand to four-thousand feet above ground level that usually flows from south to north, bringing moisture from the Gulf of Mexico.

Low-precipitation supercell. Thunderstorm with very little rain falling to the ground. Sometimes produces large hail and tornadoes.

Low-pressure area. Cyclone. In northern hemisphere, winds flowing around a central point in a counterclockwise manner. An area where the net movement of air is upward.

Managing editor. Individual responsible for news story selection.

Mesocyclone. Circulation inside a thunderstorm that meets designated parameters. Also known as the tornado cyclone.

Meteorologist. Individual involved in the study and/or forecasting of the weather. Requires a university degree in meteorology or a degree in math or physics with the appropriate meteorology classes.

Microwave truck. Vehicle equipped with electronics that allow line-of-sight transmission of live television video and audio.

Multiple vortex. Refers to the situation in which a mesocyclone produces several tornadoes at the same time.

News director. Individual responsible for entire news operation.

One-hundred-year storm. A storm that is expected to occur at a given location at least once every one hundred years.

Outflow. Horizontal winds resulting from a thunderstorm downdraft.

Overhang. Radar feature that is one of the parameters that helps to designate a thunderstorm as a supercell.

Overrunning. Usually describes a situation where warm, moist air flows up over colder, more dense air.

Precipitation. Liquid or frozen drops of moisture that fall from the atmosphere.

Producer. Individual directly responsible for preparing and overseeing newscasts.

Radar. Acronym derived from *ra*dio *d*etecting *a*nd *r*anging. Used in meteorology to detect precipitation intensity and speed of movement.

Relative humidity. The amount of moisture in the air relative to how much it can hold at a given temperature and pressure.

Seabees. Navy term for personnel involved in construction.

Squall line. A fairly narrow line of thunderstorms.

Stable. Refers to an atmospheric condition where upward motion of the air is suppressed.

Station manager. The individual who is usually second in authority at a television station.

Storm surge. Ocean water that increases in height and moves inland with a hurricane.

Supercell. A thunderstorm that exhibits certain well-defined characteristics and usually produces severe weather.

Temperature. The measure of the degree of thermal or heat intensity of air, water, and so on.

Tornado. A violently rotating column of air in contact with the ground.

Tornado cyclone. Mesocyclone inside the thunderstorm that may produce a tornado.

Tornado vortex signature (TVS). Radar presentation that meets designated parameters. Is the actual tornado circulation.

Tornado warning. Issued when a tornado has been sighted or is indicated on radar. Covers a small area and is usually in effect for an hour or less.

Tornado watch. Issued when there is a chance of tornadoes. Covers hundreds of square miles and is usually in effect for several hours.

Tropical air. Warm, moist air that originates over tropical ocean areas.

Trough. An elongated area of atmospheric low pressure.

Twister. A tornado.

Unstable. Refers to an atmospheric condition in which upward motion of air tends to continue until the temperature of the rising air is the same as that of the surrounding air.

Vortex. Used to describe a tornado. Also the center of circulation of any low-pressure area.

Wave force. The pressure exerted by ocean waves. A result of water density and speed of wave movement.

Warm front. A zone of transition where warmer air is replacing colder air.

Index

Television personnel: drug problems, 76–77; journalists and photographers, 76, 80; lightning strike of engineer, 190–91; shooting of photographer, 183–84; training of, 182; weather department employees, 108–109. *See also* News anchors; News directors

Television weather, 50; behind the scenes, 142–43; "blue sky" reports, 189–90; false weather reports, 181–82, 185, 190, 192–94; forecasting, 16–19; KWTV weather department, 197; meteorologists, 201, 202; meteorologists, criticism of, 75–76; meteorologist's work schedule, 179; new employees, 108; operational hierarchy, 54; performance standards, 109; promotion of personalities, 85; rating wars, 181, 185, 187–89, 192–95; responsibility of, 139

Temperature, 28; and tornadoes, 89, 91

Terns, 28

Texas, 19, 20, 111–13, 200

Thomas, Bob, 34–35

Those Terrible Twister shows, 150–53

Thunder, 176

"Thunder lizard," 49, 57–59

Thunder snow, 90

Thunderstorm Project, 20

Thunderstorms, 16, 20, 90; folk predictions, 11; lightning and thunder, 175–76; with mesocyclones, 72, 163, 176; radar probing of, 69, 71; tornadoes and, 12, 68, 99

Tinker Air Force Base, Midwest City, Okla., 14–15, 22, 23

Tomez, Dave, 112

Tornadoes, 14, 139, (1905) 8, (1947) 4–5, 8, (1948) 14, 15, (1953) 19, (1955) 20, (1956) 20, (1970) 42–43, (1973) 66–71, 73, 77–78, (1974) 79–84, (1975) 90, 91, (1977) 98–101, (1978) 103, (1979) 111, 112, 113, (1980) 116, (1981) 119, 120–23, 126–34, (1982) 137–38, 147, (1984) 157–59, (1985) 161, (1986) 161–72, (1988) 174, (1990) 182–83, 186–87, (1991) 188, (1992) 190–91, (1993) 193, 194, (1994) 199, 200, 201; forecasting, 14–15, 17; generation, 98; largest recorded outbreak, 79; live video, 127–28; mesocyclone rotation and, 174; radar probing of, 71; safety precautions, 10, 169–70; spin, direction of, 94; temperatures and dew points, 89, 91; vortex pattern, 72; weather wars, 181–82, 185, 187, 192. *See also* Fatalities, tornado

Tornado vortex signature (TVS), 72, 130

Tornado warnings, 12, 13, 20, 66, 67; automatic warning system, 188; first real-time Doppler radar tornado warning, 137–39; priority one, 81; television warning vs. official warning, 139, 140, 141; Union City tornado (1973), 69, 70